2400

Scientific Inference,
Data Analysis, and Robustness

Publication No. 48
of the Mathematics Research Center
The University of Wisconsin—Madison

Academic Press Rapid Manuscript Reproduction

Scientific Inference, Data Analysis, and Robustness

Edited by

G. E. P. Box
Tom Leonard
Chien-Fu Wu

Mathematics Research Center and Department of Statistics
The University of Wisconsin — Madison
Madison, Wisconsin

Proceedings of a Conference Conducted by the
Mathematics Research Center
The University of Wisconsin—Madison
November 4–6, 1981

1983

ACADEMIC PRESS
A Subsidiary of Harcourt Brace Jovanovich, Publishers

New York   London
Paris   San Diego   San Francisco   São Paulo   Sydney   Tokyo   Toronto

This work relates to Department of the Navy Grant N00014-82-G-0008 issued by the Office of Naval Research. The United States Government has a royalty-free license throughout the world in all copyrightable material contained herein.

ACADEMIC PRESS, INC.
111 Fifth Avenue, New York, New York 10003

United Kingdom Edition published by
ACADEMIC PRESS, INC. (LONDON) LTD.
24/28 Oval Road, London NW1 7DX

Library of Congress Cataloging in Publication Data
Main entry under title:

Scientific inference, data analysis, and robustness.

(Publication no. 48 of the Mathematics Research
Center, The University of Wisconsin--Madison)
"Proceedings of the Conference on Scientific
Inference, Data Analysis and Robustness ... sponsored
by the United States Army under Contract
no. DAAG29-80-C-0041"--
Includes index.
1. Mathematical statistics--Congresses. 2. Robust
statistics--Congresses. I. Box, George E. P.
II. Leonard, Tom. III. Wu, Chien-Fu. IV. University
of Wisconsin--Madison. Mathematics Research Center.
V. Conference on Scientific Inference, Data Analysis,
and Robustness (1981, Madison, Wis.) VI. United
States. Army. VII. Series: Publication of the
Mathematics Research Center, the University of
Wisconsin--Madison ; no. 48.
QA3.U45 no. 48 [QA276.A1] 510s [519.5] 82-22755
ISBN 0-12-121160-6

Contents

Contents

Contributors

Numbers in parentheses indicate the pages on which the authors' contributions begin.

Hirotugu Akaike (165), The Institute of Statistical Mathematics, 4-6-7 Minami-Azabu, Minato-ku, Tokyo 106, Japan

David F. Andrews (153), Department of Statistics, University of Toronto, Toronto, Ontario, Canada M5S 1A1

G. A. Barnard (1), Faculty of Mathematics, Department of Statistics, University of Waterloo, Waterloo, Ontario, Canada, N2L 3G1

G. E. P. Box (51), Mathematics Research Center and Department of Statistics, University of Wisconsin—Madison, Madison, Wisconsin 53706

A. P. Dempster (117), Department of Statistics, Harvard University, Cambridge, Massachusetts 02138

L. Y. Deng (245), Department of Statistics, University of Wisconsin—Madison, Madison, Wisconsin 53706

Persi Diaconis (105), Department of Statistics, Stanford University, Stanford, California 94305

David Freedman (105), Department of Statistics, University of California at Berkeley, Berkeley, California 94720

I. J. Good (191), Department of Statistics, Virginia Polytechnic Institute and State University, Blacksburg, Virginia 24061

David V. Hinkley (85), School of Statistics, University of Minnesota, Minneapolis, Minnesota 55455

J. Kiefer* (279), Department of Statistics, University of California at Berkeley, Berkeley, California 94720

Tom Leonard (9), Mathematics Research Center and Department of Statistics, University of Wisconsin—Madison, Madison, Wisconsin 53706

C. L. Mallows (135), Bell Laboratories, Murray Hill, New Jersey 07974

Carl N. Morris (25), Department of Mathematics, University of Texas at Austin, Austin, Texas 78712

Donald B. Rubin (213), Mathematics Research Center, University of Wisconsin—Madison, Madison, Wisconsin 53706

C. F. Jeff Wu (245), Mathematics Research Center and Department of Statistics, University of Wisconsin—Madison, Madison, Wisconsin 53706

H. P. Wynn (279), Department of Mathematics, Imperial College, South Kensington, London S.W. 7, England

*Deceased

Foreword

This volume contains the Proceedings of the Conference on Scientific Inference, Data Analysis, and Robustness held in Madison, Wisconsin on November 4–6, 1981, under the auspices of the Mathematics Research Center, University of Wisconsin—Madison. The conference was sponsored by the United States Army under Contract No. DAAG29-80-C-0041 and supported by the Office of Naval Research, U.S. Navy, under Grant N00014-82-G-0008. The five sessions were chaired by: Irwin Guttman, University of Toronto; George Tiao, University of Wisconsin—Madison; Robert Miller, University of Wisconsin—Madison; Morris DeGroot, Carnegie-Mellon University; George Box, University of Wisconsin—Madison. There were about 200 participants, and in addition to the speakers and chairmen, invited participants included Ching-Shui Cheng, Seymour Geisser, David Hinkley, Robert Hogg, Keith Ord, and Stephen Stigler, whose after dinner speech on the history of Bayes theorem was a highlight of the conference. The conference was preceded by a two-day short course on Bayesian Inference and Modelling.

The program committee consisted of George Box, Tom Leonard, and C. F. Jeff Wu. Mrs. Gladys Moran handled the administration of the conference.

We are much indebted to Elaine DuCharme who with great care and despatch has put this volume together and compiled the index.

Preface

The statistical profession has in recent years been divided into frequentists, Bayesians, likelihood adherents, fiducialists, structuralists, and so on. Much of the controversy has surrounded inferences about unknown parameters, conditional on the truth of the sampling model. Such concepts have been invoked as the likelihood principle, coherence, and frequentist optimality, given the model. However, because in practical situations the model is never precisely specifiable, these idealized concepts alone seem insufficient to the needs of scientific model building. The future of statistical methodology probably lies in close collaboration between statisticians and scientific practitioners to ensure that developments in statistical methodology genuinely address themselves to the broadening of scientific knowledge.

The main purpose of the conference was to study the philosophy of statistical modelling, including model robust inference, and the appropriate analysis of data sets. Therefore, experts with a wide range of views were invited. There was considerable support for the view that whilst Bayes and likelihood might be good for parametric inference, frequentist ideas were needed for checking models. Hence, it might be desirable for a statistician to refer to more than one theory of inference at different stages in the analysis.

The relevance of significance tests was discussed at length, and provoked lively debates, in particular after the presentations by Box, Dempster, and Leonard. The consensus seemed in favor of significance tests for model checking, but there was less agreement on tests for parameters. The questions as to how to judge whether a p-value is small enough to reject the model was raised. Some participants felt that this was not a valid practical criticism, and others thought that the statistical user needed further guidance, and should not just rely on standard conventions.

The importance of data analysis and data description in scientific modelling was highlighted in papers by Dempster, Mallows, Andrews, Efron, and Rubin. Also, computer simulations were discussed, by Efron for bootstrap techniques, and by Rubin for Bayesian posterior densities.

Morris and Good presented papers on empirical and hierarchical Bayes estimation. For linear or multinomial models, the prior parameters can frequently be estimated from the data, and the shrinkage estimators for the model parameters can smooth the classical estimators in a sensible way which seems clearly advantageous.

Diaconis described some counter examples to nonparametric Bayesian modelling, involving the use of prior distributions across function space. It was however suggested that, when the prior is put on sensibly, then a sensible posterior estimate of the sampling model will ensue. There is considerable further scope for model space priors; this seems to be one of the few ways forward for developing theoretically sound techniques for the construction, rather than the checking, of models. These possibilities were also discussed by Leonard.

Model robust techniques were discussed by Barnard, Box, Wu and Deng, and Kiefer and Wynn. Much of the robustness literature is concerned with the possibility of leptokurtosis in the error distribution; but skewness, bimodality, nonlinearity of relationships, and nonindependence within a sample deserve equal attention.

Hinkley described some interesting examples which illustrate how it may be necessary to mix mathematical techniques with intuitive reasoning. Akaike discussed the parallels between information theory and statistical modelling, and described how the concept of entropy may be used to develop a generally applicable model selection criterion.

There was, overall, a broad measure of agreement on future directions for our subject. In particular a better understanding of appropriate roles for mathematical theory, practical and intuitive reasoning, and data analysis seems to be developing.

George Box,
Tom Leonard,
and Chien-fu Wu

Pivotal Inference
and the Conditional View of Robustness
(Why have we for so long managed
with normality assumptions?)

G. A. Barnard

0. SUMMARY.

A major reason for normality assumptions has been that
standard, unconditional theories of inference require such
assumptions to avoid excessive computational problems. The
pivotal approach, like the Bayesian approach, is
conditional, and as such does not need such assumptions,
given facilities now available on hand-held computers. When
there is doubt about distributional assumptions, therefore,
a range of such assumptions can be tested for their effect
on the inferences of interest. Some samples will be robust
with respect to possible changes in distribution, while
other samples will not be. When the latter is the case, the
statistician should draw attention to the fact, in
accordance with the principle that it is at least as
important for the statistician to tell his clients what they
do not know as it is to tell them what they do know from
their data. In sampling from 'normal-looking'
distributions, such as the Cauchy, treating 'normal-looking'
samples as if they were normal produces errors unlikely to
be of practical importance. The non-robust samples are
those presenting non-normal features, such as skewness. In
the past such samples have been treated by ad hoc methods.
Statisticians should make it their business to acquire
empirical knowledge of the types of distribution to be met

1

with the areas of application with which they are
concerned. Skewness of distribution is particularly to be
watched for.

　　　1. Problems of uncertainty of distributional form
arise almost exclusively in connection with continuous
observables. For any such set $\underline{x} = (x_1, x_2, \ldots, x_n)$ a
probabilistic model will amount to asserting that there is a
parameter $\underline{\theta} = (\theta_1, \theta_2, \ldots, \theta_k)$ such that, given $\underline{\theta}$, $\underline{x}$ has
probability density $\phi(\underline{x}, \underline{\theta})$, where ϕ is approximately
known. If, from the marginal density $\phi_1(x_1; \underline{\theta})$ we derive
the probability integral transformation

$$p_1 = p_1(x_1, \underline{\theta}) = \int_{-\infty}^{x_1} \phi_1(t; \underline{\theta}) dt$$

as is well known, $p_1(x_1, \underline{\theta})$ will be uniformly distributed
between 0 and 1. Then we can form the (marginal) density
of x_2, given x_1, $\phi_{12}(x_2; \underline{\theta}, x_1)$, and thence obtain

$$p_2 = p_2(x_2, x_1, \underline{\theta}) = \int_{-\infty}^{x_2} \phi_{12}(t; \underline{\theta}, x_1) dt$$

again uniformly distributed between 0 and 1. Continuing
in this way we can find a set $p = (p_1, p_2, \ldots, p_n)$ of
functions of the $\underline{x}$ and of $\underline{\theta}$ such that, to say that $\underline{x}$,
given $\underline{\theta}$, has the density ϕ is equivalent to saying that
the specified functions $\underline{p}$ of $\underline{x}$ and $\underline{\theta}$ are uniformly
distributed in the unit cube. Following Fisher we call a
function of observations and parameters whose distribution
is known, a <u>pivotal quantity</u>, and the vector function
 $\underline{p}(\underline{x}, \underline{\theta})$ will be (a form of) the <u>basic pivotal</u> of the model
we are discussing. This mode of expressing a probability
model by means of pivotal functions uniformly distributed in
the unit cube was first put forward in 1938 by Irving
Segal. Insofar as we think of ϕ as only <u>approximately</u>
known, so the pivotal $\underline{p}$ will be only <u>approximately</u>
uniformly distributed. More generally, it may be convenient
to take the basic pivotal $\underline{p}(\underline{x}, \underline{\theta})$ to have an approximately
known distribution other than uniform -- for example, in the
cases to which, for simplicity, we restrict ourselves, where

$k = 2$ and $\underline{\theta} = (\lambda, \sigma)$, with λ a location parameter and σ a scale parameter, we can set

$$p_i = (x_i - \lambda)/\sigma \tag{1}$$

and express our information about the distributional form by saying that $\underline{p}$ has the density $f(\underline{p})$, where f might be standard normal, or standard Cauchy, or some other standard distribution. We shall consider the problem of estimating λ, regarding σ as a nuisance parameter. Our arguments can be extended without essential change to general regression models; models of still greater generality will typically require approximations into the details of which we do not enter here.

The pivotal formulation of a model has a double advantage. First, the <u>meaning</u> of the parameters θ is defined by reference to the way they enter the basic pivotal $\underline{p}$, rather than by reference to a particular feature, such as the mean, the mode or the center of symmetry, of the distribution. Thus the difficulty faced by studies such as that of Andrews et al., in which the location parameter had to be taken as the center of symmetry -- thus limiting consideration to cases where the distribution could be taken to be exactly symmetrical -- can be avoided. Second, as was first stressed by Dempster, specifying the distribution by means of pivotals enables us to reduce, if not eliminate, the sharpness of the distinction between observables and parameters; indeed, we may extend the pivotal model by saying that $\theta = (\underline{\theta}_0, \underline{\theta}_1)$ and that the basic pivotal is $(\underline{p}, \underline{\theta}_1)$, with density $f(\underline{p})\pi(\underline{\theta}_1)$, corresponding to the assertion that the part θ_1 of the parameter θ has the prior density $\pi(\theta_1)$. If the part $\underline{\theta}_0$ were empty we would then be formulating a fully Bayesian model, while if $\underline{\theta}_1$ were empty our model would be fully non-Bayesian. Since our mode of reasoning is the same, we need not commit ourselves in advance to being Bayesian or non-Bayesian.

2. The general inference procedure is to make 1-1 transformations on $(\underline{p}, \underline{\theta}_1)$ to bring it, so far as possible, to the form $(\underline{T}, \underline{N}, \underline{A})$. Here $\underline{T}$ is of the form $\underline{T}(\underline{x}, \underline{\theta}_i)$, where θ_i denotes the parameter(s) of interest, and for

each fixed value x_0 of the observations, $T(x_0, \theta_i)$ defines a 1-1 mapping between the range of T and the range of θ_i. N is similarly a function of the nuisance parameters, $N(x, \theta_n)$, while $A(x)$ does not involve the parameters. Since (T, N, A) is a 1-1 function of (p, θ_1), its density is approximately known; then, given the observations $x = x_0$, the value of A will be known, and the relevant density of (T, N) will be the conditional density, given $A(x) = A(x_0)$. Integrating N out from this density will given a density for T which can be used to derive a confidence distribution for θ_i.

This inference procedure will often not be capable of being carried through exactly, and then we must resort to approximations. But in the case of location and scale, and more generally regression problems, we can carry it through exactly, as follows: Taking p as in (1), we require

$$\partial T/\partial \sigma = 0, \quad \partial N/\partial \lambda = 0, \quad \text{and} \quad \partial A/\partial \lambda = \partial A/\partial \sigma = 0 . \qquad (2)$$

Now

$$\partial A/\partial \lambda = \sum_i (\partial A/\partial p_i)(\partial p_i/\partial \lambda) = (-1/\sigma) \sum_i \partial A/\partial p_i$$

while

$$\partial A/\partial \sigma = \sum_i (\partial A/\partial p_i)\partial p_i/\partial \sigma) = (-1/\sigma) \sum_i p_i \partial A/\partial p_i .$$

so that A must satisfy the PDE's

$$\sum_i \partial A/\partial p_i = 0, \qquad \sum_i p_i \partial A/\partial p_i = 0 \qquad (3)$$

the general solution to which is an arbitrary function of the $n - 2$ functionally independent quantitites $c_1, c_2, \ldots, c_{n-2}$, where

$$c_i = (p_i - \bar{p})/s_p = (x_i - \bar{x})/s_x, \quad i = 1, 2, \ldots, n \qquad (4)$$

and $\bar{x}$ and s_x are used in the customary way to denote the mean and S.D. of a finite set of quantities. Since we clearly want the maximal possible conditioning, we take $A = c$. We could then use standard methods of the theory of PDE's to arrive at expressions for T and N, but it is obvious that the 1-1 transformation

$$p_i = N(T + c_i) \tag{5}$$

will serve our purpose; for since $\sum_i c_i = 0$, summing over i gives

$$\bar{p} = NT, \quad p_i - \bar{p} = Nc_i, \quad \text{so} \quad N = s_p = s_x/\sigma \tag{6}$$

and finally $T = (\bar{x} - \lambda)/s_x$. The functions N and T are clearly of the form required.

The transformation (5) can be shown to have Jacobian

$$\sqrt{n}(n - 1)(n - 2)N^{n-1}/|c_n - c_{n-1}| \tag{7}$$

so that the joint density of $(T, N, \underline{A})$ is

$$\sqrt{n}(n - 1)(n - 2)N^{n-1}/|c_n - c_{n-1}| \cdot f(N(t \cdot \underline{1} + \underline{c})) \tag{8}$$

Given the observations, the value of $\underline{c} = \underline{c}_0$ is known, so that joint conditional density of (T, N) is

$$KN^{n-1}f(N(T \cdot \underline{1} + \underline{c}_0))$$

and the marginal density of T is

$$\xi(T; \underline{c}) = K \int_0^\infty z^{n-1} f(z(T \cdot \underline{1} + \underline{c}_0)) dz . \tag{9}$$

where K is determined by the condition that ξ integrates to 1.

Provided $(\bar{x} - \lambda)\sqrt{n}/s_x = t$ has a non-singular distribution, as will very often be the case, we can change the variable in (9) from T to t, and put $u = zt$, $dz = du/t$ in the integral to obtain

$$\xi^x(t; \underline{c}) = (K*/t^n) \int_0^\infty u^{n-1} f(u(\underline{1} + (1/t)\underline{c}_0)) du \tag{10}$$

which is $O(1/t^n)$ as $t \to \infty$. Thus the tail behaviour of the conditional density of t is in this broad class of cases, the same as in the case of normality. This is one reason why interpretations of Student's t as if the original density was normal have often been adequate in practice.

3.   It is instructive to study the case when the observations are independent and from a Cauchy distribution. The algebra becomes heavy for n greater than 5 but, for example for n = 3 we have

$$f(\underline{p}) = 1/\pi^3(1 + p_1^2)(1 + p_2^2)(1 + p_3^2)$$

and putting $b_i = t + c_i$ in (10) we get

$$\xi^X(t;\underline{c}) = -K^{**} \int_0^\infty z^2 dz / \prod_i (b_i z - i) \prod_i (-b_i z - i)$$

$$= -K^{**} \int_0^\infty z^2 dz / h_3(z) h_3(-z)$$

and this can be evaluated by formulae given in Gradshtyn and Ruzhik (p. 218). We find

$$\xi^X(t;\underline{c}) = K^{**}/(S_1 S_2 - S_3)$$

where S_k stand for the sum of the products, k at a time, of the quantities $b_i = t + c_i$. Plots of $\xi^X(t;\underline{c})$ for various configurations $\underline{c}$ are easily produced using an IIP41C with plotter. Comparison with the plot of the density of the normal Student's t with 2 degrees of freedom shows that for the symmetric configuration $\underline{c} = (-\sqrt{3}, 0, +\sqrt{3})$ there will be little error in using the normal tables beyond the values t = ±2.2 -- and, of course it is this range which will be the one most frequently used. Things are otherwise with the skew configurations, especially with the extremely skew $\underline{c} = (-1.4908, -0.4092, +1.9000)$; but even here, provided one is concerned with two-sided prob- abilities, and with sufficiently large values of t, the errors involved are not particularly serious.

4.   Another density worthy of study is the Barndorff- Nielsen density with

$$\ln f(\underline{p}) = \sum_i (K - \sqrt{(1 + p_i^2)} - \beta p_i)$$

where β is a skewness parameter. When $\beta = 0$ this density approximates the standard normal in the center of its range, but in the tails it goes down as e^{-p_i} instead of as $e^{-p_1^2}$. It appears that use of normal tables here does little harm provided whatever skewness there is in the sample fairly reflects the skewness in the true distribution; the worst case arises when there is skewness in the sample in a sense opposite to that in the true distribution.

Whenever the density is such that

$$\ln f(\underline{p}) = K - H(\underline{p})$$

where $H(\underline{p})$ is homogeneous of degree α, so that $H(z\underline{p}) = z^{\alpha}H(\underline{p})$ it is possible to evaluate the integral (10) in closed form. Because then

$$\xi^*(t;\underline{c}) = K^* \int_0^{\infty} z^{n-1}\exp -H(z(T\cdot\underline{1} + \underline{c}_0))dz$$

$$= K^* \int_0^{\infty} z^{n-1}\exp -z^{\alpha}L\cdot dz$$

where

$$L = H(T\cdot\underline{1} + \underline{c}_0) .$$

The integral becomes a gamma integral by the substitution $v = z^{\alpha}L$ and we find

$$\xi^*(t;\underline{c}) = K^{**}/(H(T\cdot\underline{1} + \underline{c}_0))^{n/\alpha} .$$

The independent or dependent normal distribution is a special case of this. In particular, for the independent normal, $H(\underline{p}) = \frac{1}{2}\underline{p}'\underline{p}$, $\alpha = 2$, and

$$H(T\cdot\underline{1} + \underline{c}_0) = (T\cdot\underline{1} + \underline{c}_0)'(T\cdot\underline{1} + \underline{c}_0)$$

$$= (nT^2 + \underline{c}_0'\underline{c})$$

$$= (t^2 + (n - 1))$$

giving the usual result for Student's t-density. In the case of the double exponential, $H(p) = \sum_i |p_i|$ and $\alpha = 1$. It will be noted that for the normal distribution, the

resultant t density does not involve $\underline{c}$, so in this case
-- and essentially only in this case together with that of
the uniform density -- the conditional approach and the
marginal approach give the same result.

   5. There exists a wide area of extremely useful
research to be done in finding which non-normal densities
apply to which types of empirical data. Barndorff-Nielsen
has shown that his family of hyperbolic distributions fit
very well to distributions arising in connection with
turbulent flows of various sorts. Karl Pearson and Weldon
and their collaborators --including Student himself -- did
great work early in the century in connection with
biometrical distributions. But for the past fifty years
work of this kind has been neglected -- presumably because
little was known of how to use such information, and the
computing facilities needed, now available on quite small
computers, was simply not available.

   For those whose limited access to sets of empirical
data prevent them from engaging in the useful research
indicated, there is the purely theoretical problem still, so
far as I know, without a solution -- how far we can
determine the form of a density, given an arbitrarily large
number of samples of a fixed finite size, with varying and
unknown location and scale parameters.

                         Faculty of Mathematics
                         Department of Statistics
                         University of Waterloo
                         Waterloo, Ontario, Canada
                         N2L 3G1

Some Philosophies
of Inference and Modelling

Tom Leonard

SECTION 1:   INTRODUCTION.

During the Spring semester of 1981, the Mathematics
Research Center held a weekly statistical discussion series
as a precursor to its special year on Scientific Inference,
Data Analysis, and Robustness. The many discussants
included G. E. P. Box, D. V. Lindley, B. W. Silverman, A.
Herzberg, C. F. Wu, B. Joiner and D. Rubin. Many aspects of
statistics were discussed, including the Box philosophy of
deductive and inductive reasoning, and Lindley's coherent
Bayesian viewpoint. The present paper attempts to
constructively review the discussion series, and to add a
number of retrospective comments and suggestions.

SECTION 2:   SESSIONS 1 TO 3 WITH FURTHER IDEAS ON MODELLING,
             SIGNIFICANCE TESTING AND SCIENTIFIC DISCOVERY.

Session 1:   Checking Models, George Box (1/23/81)

In the first session the deductive and inductive
aspects of statistical investigation were discussed.
Deduction is appropriate for inferences upon the truth of
the model, whilst inductive thought is necessary during
model checking. During the semester it became apparent that
all serious discussants were in agreement on this issue

There was a bit less agreement on which philosophy
should be employed during the model checking procedure.

Discussants seemed to split into the following three main areas:

(a) Bayes is good for inferences given the model but frequentist procedures, e.g. significance tests, are necessary when checking the model.

(b) Bayes is good for inferences given the model, and Bayes is also good for model-checking (e.g. prior distributions on either sampling densities or different models or polynomial coefficients) but more Bayesian theory needs to be developed in the model-checking area.

(c) Frequentist procedures are adequate for both inferences and model-checking.

The main debates were between (a) and (b). Frequentist model-checking needs few assumptions about alternative models, whilst Bayesian assumptions always reduce to a grand model involving models across models. Therefore frequentist model-checkers can point to the simplicity and generality of their approach, whilst Bayesians could give the response that it is always necessary to inject a certain amount of structure into the analysis in order to focus upon precise conclusions. This is an important issue which was largely unresolved.

The Inductive Modelling Process (IMP) and the less critical importance of coherence were discussed, with interesting responses by D. V. Lindley and A. F. M. Smith, by Leonard (1981a)

Session 2: The Likelihood Principle, Tom Leonard (1/30/81)

In the second session the Likelihood Principle was introduced in the context of making inferences conditional upon the truth of the model, and the proof of Birnbaum's theorem was presented. This proves that if the statistician accepts the sufficiency and conditionality principles (which are open to straightforward frequentist interpretations) then he must accept the Likelihood Principle, conditional upon the truth of the model, and should not therefore employ any approach involving integrations across the sample space (e.g. UMVU estimation, confidence intervals, significance tests).

The reaction to these ideas was interesting. Bayesians viewed the sufficiency and conditionality principles as obviously acceptable. Traditional significance testers felt that, since the Likelihood principle and testing are not compatible, there must be something misleading in the underlying assumptions (most likely the Conditionality principle). Another expressed view was that the Likelihood Principle is largely irrelevant since it conditions on the truth of the model, whilst most of the statistician's effort needs to be spent on model-building. One nice interpretation was that "if your analysis does not satisfy the Likelihood Principle than this means that your model is wrong".

Overall, it seemed that few existing views were changed by this exposure to Birnbaum's theorem. This may be viewed as surprising, as the proof and underlying assumptions for this theorem are extremely plausible and simple. (They seem to be a bit more plausible than the rather tautologous assumptions underlying the concept of "coherence" - see Leonard (1981a)).

Session 3: Tying together the ideas of the last two
            sessions, Dennis Lindley,

In the third session an attempt was made to extend the Likelihood Principle ideas from the inferential to the modelling situation. This involved a prior distribution across model space, and the ideas therefore needed to be partly interpreted from a Bayesian point of view. The main reactions were either (a) a Likelihood Principle would be neither reasonable nor desirable in modelling situations since frequentist ideas are obviously more appropriate on model space, or (b) these ideas would be desirable in modelling situations but some further theoretical development would be needed in order to obtain a modelling principle with the same impact for non-Bayesians as the Likelihood Principle for inference.

Perhaps the discussants in Session 3 might have favourably considered the following principle:

The Modelling Principle (special case of Sufficiency Principle and of the Likelihood Principle)

Suppose that the outcome of an experiment is the numerical realization $\underset{\sim}{x}$ taking values in a sample space H. Let $f_1(\cdot)$ and $f_2(\cdot)$ be two probabiity densities defined on H (with respect to an appropriate dominating measure) such that

$$f_1(\underset{\sim}{x}) = f_2(\underset{\sim}{x})$$

Then unless there is information external to the data to suggest otherwise, neither of f_1 and f_2 should be viewed as preferable for modelling conclusions based on $\underset{\sim}{x}$.

N.B. For each i, $f_i(\underset{\sim}{x})$ is a probability density, conditional on the unknown "parameter" f_1. This may also be interpreted as the likelihood functional of f_i conditional on $\underset{\sim}{x}$. The Modelling Principle is saying that if f_1 and f_2 posses the same likelihood functional then they should be viewed as equally preferable for modelling conclusions based on $\underset{\sim}{x}$, in the absence of external information.

The Modelling Principle could be used to critically interpret well-known modelling approaches due to Tukey and Parzen. Whilst Box's modelling approach is particularly well-formulated; the following example is interesting. Note that problems with events of probability zero, whilst unimportant, could be removed by extending the Modelling Principle to say that, in the absence of external information, f_1 should be preferred to f_2 whenever
$f_1(\underset{\sim}{x}) > f_2(\underset{\sim}{x})$.

Example - Box's Modelling Approach

Consider an observation vector $\underset{\sim}{x} = (x_1, \ldots, x_n)^T$ assuming values in the sample space Ω which we take to be n-dimensional Euclidean space R^n. Suppose that $g(\cdot) : R \rightarrow R$ is some monotonic transformation on the real line (e.g. a Box-Cox transformation). Assume further that the observed elements of $\underset{\sim}{x}$ happen to satisfy the specific condition

$$\sum x_i^2 = \sum g^2(x_i) - 2 \sum \log \left| \frac{\partial g(x_i)}{\partial x_i} \right| = s^2 \tag{$*$}$$

Consider the alternative models M_1 and M_2 specified by
M_1: The x_i are realizations of independent random
variables x_i which possess standard normal distributions.
The corresponding probability density is

$$f_1(\underset{\sim}{x}) = \frac{1}{(2\pi)^{1/2} n} \exp\{-\frac{1}{2} \sum_i x_i^2\} \quad \text{for} \quad \underset{\sim}{x} \in \Omega$$

M_2: The x_i are realizations of independent random
variables X_i such that the transformed variables
$Y_i = g(X_i)$ possess standard normal distributions. The
probability density is now given by

$$f_2(\underset{\sim}{x}) = \frac{1}{(2\pi)^{1/2} n} \exp\{-\frac{1}{2} \sum_i g^2(x_i) + \sum_i \log |\frac{\partial g(x_i)}{\partial x_i}|\}$$

$$\text{for} \quad \underset{\sim}{x} \in \Omega \ .$$

Note that under condition (*) $f_1(\underset{\sim}{x}) = f_2(\underset{\sim}{x})$ so that
whenever $\underset{\sim}{x}$ satisfies (*), the Modelling Principle tells us
to prefer f_1 and f_2 equally in the absence of external
information.

The Box modelling approach tells us to discredit M_1
if the tail probability

$$\theta_1 = p(f_1(\underset{\sim}{X}) < f_1(\underset{\sim}{x}))$$

$$= p(\sum_i x_i^2 > s^2)$$

is to small, where the probability on the right hand side
arises from the distribution of $\underset{\sim}{X}$ under M_1, and s^2 is
specified in (*). Note that θ_1 is just the probability
that a chi-squared random variable, with n degrees of
freedom is greater than or equal to s^2.

We should also discredit M_2 if

$$\theta_2 = p(f_2(\underset{\sim}{X}) < f_2(\underset{\sim}{x}))$$

$$= p(\sum_i x_i^2 - 2 \sum_i \log |\frac{\partial g(X_i)}{\partial X_i}| > s^2)$$

is to small, where the probability on the right hand side is
now based upon the distribution of $\underset{\sim}{X}$ under M_2. Since
$\sum_i Y_i^2$ possesses a chi-squared distribution with n degrees
of freedom, and this is adjusted by an extra function of

$\underset{\sim}{X}$, we see that θ_1 will not in general be the same as θ_2. Therefore, although the Modelling Principle tells us to equally prefer M_1 and M_2, we seem to arrive at different tail probabilities in each case.

Our overall conclusion is that Box's Modelling Approach and the Modelling Principle are not in mathematical agreement. This may be the source of some discussion. Perhaps the philosophy "all principles are there to be broken, but in this case we may learn a great deal by considering whey we have broken them" may be useful here.

Significance Testing

During the first three sessions a large amount of time was spent discussing the merits of significance testing. There was some measure of agreement that these are unreasonable given the model but much less agreement in the modelling situation. The main points raised by objectors to significance tests were

(a) Fixed size tests are fairly arbitrary and it seems to be extremely difficult to interpret the magnitude of the p-value when so many different aspects like sample size, model complexity, and selective reporting affect the p-value.

(b) There is no justification for making accept/reject decisions based on significance tests.

(c) It is dangerous to summarize the results of an experiment by a single p-value.

(d) In modelling situations it is necessary to have alternatives in mind; standard tests for fit do not involve alternative models and may therefore not be based upon enough assumptions to facilitate useful conclusions.

Proponents of significance tests made the following points:

(a) The p-value can be interpreted very naturally either by thinking in terms of the tail area of the sampling distribution or by comparison with the p-values of other experiments. Interpretations based upon surprise factors are particularly important.

(b) When the majority of effort is spent on model-building it then seems rather unimportant to argue about the difference between 5% and 4% at the end of the analysis.

(c) The p-value is only one of a large number of aspects which a statistician should think about in reaching his conclusion. It is not a formal mechanism e.g. for decision making, but simply a valuable guide to the inductive thought processes.

(d) When checking a model it is impossible to have all possible alternatives in mind, and therefore any procedure which conditions upon alternative models must be inadequate, thus, for example, ruling out any Bayesian procedure.

In summary, whilst tests for fit might be viewed as more appropriate than tests for parameters within a model, the big question is whether or not they indeed produce the goods i.e. do they provide a completely acceptable procedure for model-checking in the absence of alternative hypotheses, or is more structure needed in order to arrive at really convincing conclusions? In other words, can p-values for tests for fit be interpreted in a meaningful way, or is it simply too ambitious to hope to check a model unconditionally upon possible alternatives? My personal opinion would be a bit on the negative side but I would be prepared to be convinced either way. I challenge significance testers present to advise me how they actually make a practical judgement about a p-value; I remain unconvinced that they do much more than think in terms of 1% and 5%.

DISCOVERY AND INSIGHT AS OBJECTIVES OF THE SCIENTIFIC METHOD

A primary purpose of statistics is to discover new real-life conclusions e.g. a possible association between important medical factors, new chemical components useful in, say, agriculture, or novel ways of stimulating the economy. Statistics also plays a partly confirmatory role, but this is secondary to discovery. I view insight as closely related to discovery, and insight and discovery are perhaps of equal importance. Inductive modelling combined with local deduction takes statistics out of the unreasonable restrictiveness of the Neyman-Pearson and coherent Bayesian areas, and into the forefront of science, as an important vehicle for insight and discovery.

Professor Box prefers a Bayes/frequentist compromise as a means of describing his deductive and inductive reasoning. I prefer a pragmatic Bayes/pragmatic Bayes compromise. I would for example always try to work with at least the conceptual background of a prior distribution across the space of sampling models, and perhaps to employ a pragmatic short-cut to approximate to a full blown non-parametric Bayesian procedure. For example, Schwarz's criterion provides an excellent pragmatic method for judging the degree of a polynomial approximation to a non-parametrised regression function or sampling density. In short, I have developed my own pragmatic Bayes/non-parametric Bayes procedures for coping with modelling situations, and these will be reported in detail elsewhere. It is for example possible for the statistician to introduce a hypothesized model as prior estimate, and then to let the data help him to find possible deviations from his hypothesized model. This ties in well with the deductive/inductive scheme. For fuller details of Bayesian modelling procedures see Leonard (1978, 1981b, 1982a,b,c)) and Atilgan and Leonard (1982).

Unlike Professor Box, I do not view the prior distribution of the parameters as part of the sampling model. Under a Bayesian non-parametric procedure there is no restriction to the type of sampling model which can be considered or to the type of discovery which can be made. I however find Professor Box's frequentist compromise to be of potential importance both in stimulating tremendous input into the modelling area, and in suggesting that we should check the reasonability of the prior (e.g. in its tails) as well as the reasonability of the sampling model. These are of course two separate problems. Perhaps the prior should be checked by investigating the properties of the estimates it leads to.

SECTION 3: REVIEW OF SESSIONS 4-13
Session 4, Some Thoughts on Data Analysis,
       Bernard Silverman (4/13/81)

   In the fourth session the presentation of statistical
data was discussed, and a method based upon kernel
estimators was proposed for representing a random sample by
a smooth curve. Whilst this provides concise
representations, some of the information in the sample will
be lost. There was a debate about the merits of kernels and
histograms, with histograms gaining a slight advantage.

   During this session there was also a debate about
whether anyone had ever actually analysed a random sample.
The consensus of opinion seemed to be that whilst some
random samples have at times occurred in designed
experiments, most samples have non-random characteristics.

Session 5, Randomisation, Jeff Wu (2/20/81)

   In the fifth session the merits of randomisation were
debated. It seemed to be the general opinion of both
Bayesians and frequentists that randomisation is an
invaluable device. It for example removes bias due to
factors which would be difficult to model precisely, and
also helps the statistician to cope with the problem of the
lurking variable.

   The only point of debate was whether the analysis
should be carried out conditionally or unconditionally upon
the actual design employed. This issue parallels the debate
on the Likelihood and Conditionality Principles.

   The problems of how to hunt out lurking variables, or
how to analyse data in the presence of lurking variables is
one of the most important real issues which statisticians
are faced with, particularly when analysing, say, medical or
economic data, rather than data from designed experiments.
It should probably receive much more attention than, say,
the frequentist/Bayes philosophy.

Session 6, Robust Designs, Agnes Herzberg (2/27/81)

   In the sixth session robust designs were discussed with
emphasis on the criterion of D-optimality. It was generally
agreed that

(a) The theory of experimental design should always be mixed with practical common sense, and that a pragmatic design is often more useful than a theoretically optimal design, particularly when model inadequacies are taken into account.

(b) That the criterion of D-optimality is just one way of summarizing the elements of the X^TX matrix based upon the X matrix for the assumed true model, so that D-optimal designs should be treated with a great deal of caution. Recent work by Toby Mitchell and C. F. Wu on robustification of designs may also be useful here.

Session 7, The Frontiers of Statistical Analysis,
         Brian Joiner (3/6/81)

In the seventh session the main point discussed was whether it is useful to discuss slight differences between statistical methodologies when the most serious problem with large data sets is whether they have been collected properly, or stored properly on the computer, or whether it is possible to obtain convenient summaries of the data set for a preliminary analysis. A number of data sets were presented in order to illustrate various pitfalls that may be caused by careless data-handling.

There seem to be two separate problems here; clearly data handling merits considerable attention particularly when 90% of any statistical analysis should involve careful consideration of the data, for example using scatter plots and cross-tabulations. However, having done this we still need a decent formal analysis in order to sort out the statistical variation in the data. So good data handling and good statistical methodology are both of essential importance.

There seems to be some doubt as to the wisdom of collecting large quantities of badly handled data, when only a small proportion of it may ever get analysed. Perhaps the philosophy "The greater the amount of information the less you know", is not completely out of place here.

Session 8. Subjective Probability for Data Problems,
Jim Dickey (3/13/81)

In the eighth session we switched to subjective
probabiity for no data problems; and discussed the
elicitation of prior distributions from non-statistical
experts. This is a growing area amongst many traditional
Bayesians, and there has been some progress for single
parameter problems. However, severe difficulties are faced
in multi-parameter situations because of the problem of
quantifying the possibly nonlinear interdependencies between
different parameters. So the problem shows some capability
of solution, but needs considerable more development.
Procedures suggested for ensuring coherence don't always
seem to be completely coherent themselves e.g. there is
often a heavy dependence on least squares.

Session 9. Education in Statistics, Conrad Fung (3/27/81)

A number of points relating to education in Statistics
were discussed in the ninth session. It was for example
felt that statistical teaching should relate both to current
applications of our methods and to the future careers of our
students e.g. in industry.

This seems to be of considerable importance because the
statistics we are teaching now is the statistics which is
going to be applied in industry, maybe for the next forty
years. Perhaps we need a moratorium on all "bad"
statistical methods (confidence intervals, and UMP tests?),
so that only "good" methods (pragmatic Bayes?) survive into
the next century.

Session 10, Sequential Analysis, Connie Shapiro (4/3/81)

In the tenth session the theory and practical relevance
of sequential methods were discussed. The applicability of
the Likelihood Principle was debated in the context of the
variety of stopping rules available. Another important
point is that, whilst an optimal Bayes solution is always
available, the extensive analysis may be extremely
computationally complicated so that only approximate rules
are feasible. Also, in practical situations it is generally
infeasible to make the assumptions necessary for sequential
analysis, and a pragmatic rule will often work better.

Furthermore, the advantages in using a sequential rule may be diminished when model inadequacy is taken into account.

Session 11, The Truth About Bayesian Inference,
> Steve Stigler (4/10/81)

In the eleventh session, the feasibility of judging prior opinions via the predictive distribution was discussed, with historical references to Rev. Thomas Bayer' original paper. It was suggested that a serious difficulty is caused for Bayes because, for a given sampling distribution, there may be no prior distribution corresponding to the predictive distribution selected. However this simply means that the predictive and sampling distributions have not been chosen sensibly, and therefore provides a coherency check.

Session 12, Rounding Errors in Regression,
> Don Rubin (4/13/81)

In the twelfth session we discussed an asymptotic Bayes method for rounding errors which makes opposite adjustments to those suggested by numerical analysis. This is because the posterior distribution of the rounding errors is not locally uniform since it incorporates knowledge of the regression line. This is an excellent example of a situation where Bayes and pragmatism can be mixed to good effect.

Session 13, Exchangeability in Statistics,
> Dennis Lindley (4/24/81)

In the thirteenth session we discussed the idea of conditional exchangeability of observations as a Bayesian method for interpreting data. This for example leads to a resolution of Simpson's paradox. It also highlights the Bayesian theme that it is necessary to utilize information concerning the background of the data (e.g. when deciding which factor to condition on) if we are to have any hope of drawing meaningful conclusions from a finite number of observations.

SECTION 4:   REVIEW OF CLOSING SESSION

Session 14, A Review Session of the Bull Sessions,
            Tom Leonard (5/3/81)

My overall feeling is that an ideal statistician (a)
relies on his common sense and pragmatic judgement, (b) gets
involved in the scientific background of the data, (c) is
prepared to use theory when it is likely to help him reach a
useful conclusion, (d) is unwilling to accept any
theoretical procedure unless he is convinced that it is
practically relevant, (e) is at least partly Bayesian.

I would like to predict that in the next century
statisticians will be one-third Bayesian, one-third data
analyst, and one-third scientist, i.e. they will view
statistical theory and practice and scientific background as
a single entity.

Postscript: The statistical discussion series recommenced
in the Fall of 1981, and new ideas were presented on the
topics discussed above.   The next six talks were:

"Some Approaches to Modelling" by Tom Leonard

"Time Series and Outliers" by George Tiao

"The Boundaries of Statistics" by Bob Miller

"Box's Modelling Approach for Bayes-Stein Problems" by
Kevin Little.

"The Analysis of Finite Populations" by Jeff Wu
and

"The Analysis of Transformations Revisited; A Rebuttal"
by George Box with further talks planned by Don Rubin,
Ching-Shui Cheng, Dennis Cox, and Rick Nordheim

Tapes are available for all these talks, which include
many stimulating discussions together with a number of
humorous interludes.

SECTION 5:   FINAL WORD.

The most important topic at issue has been as to
whether and how a model should be checked out against the
data, with respect to a completely general alternative.   A
valid school of thought tells us that whatever procedures
have been used to construct a model, it should be finally
checked by a significant test against a general alternative,
together with other data diagnostics, to see whether the

model is compatible with the data. Another school of thought would say that there is a limit to what the data can tell us about a proposed model. While we would certainly wish that data diagnostics could help us to both construct a model and test it, this is perhaps simply asking for more than the data can give us. I believe that the richness of statistics lies in the fact that at some point in the analysis it is necessary to make model assumptions which cannot be checked out against the data but which relate either to the scientific background or to a gut feeling that "given the information in the data we are going to need to stick to some reasonable looking assumptions to provide the hope of sensible conclusions." Furthermore, it is not the model itself which should be checked, but the scientific conclusions given the model. The model is itself an artifact enabling us to think more closely about the data in relation to its scientific background. The validity of the scientific conclusions given the model can only be discussed by reference to the scientific background and to the plausibility of the intuitive thought underlying the statistical analysis; an expert judgement is called for and it would be a mistake to demand further objectivity since this is beyond the scope of statistical methodology. A significance test for a model yields a tail probability which seems to be particularly difficult to interpret in relation to the scientific background; the test moreover provides an impression of objectivity which substitutes for a clear statement of expert judgement (given a number of subjective facets). I am a strong supporter of the Box philosophy of deductive/inductive iteration between inference and modelling and think that this gains added strength when we substitute judgement for formalism at the final stage of the iterative process.

In conclusion, I see a good future for "pragmatic" Bayesians who are more concerned with the extraction of meaningful practical conclusions and the interpretation of statistical data in relation to their scientific background, and who are prepared to think more broadly than permitted by the constraints of mathematical axioms. I think that

pragmatic Bayesianism and the Box/Good/Dempster philosophy
will provide the main thrust of statistical development for
the foreseeable future.

REFERENCES

1. Atilgan T. and T. Leonard (1982). Some Penalized
   Likelihood Procedures for Smoothing Probability
   Densities. MRC Technical Report No. 2336.

2. Leonard, T. (1978). Density Estimation, stochastic
   processes, and prior information (with Discussion).
   J. Roy. Statist. Soc. B., 40, 113-146.

3. _____ (1981a). The roles of coherence and
   inductive modelling in Bayesian Statistics (with
   Discussion). In Bayesian Statistics (ed. by J.
   Bernardo), University of Valencia Press.

4. _____ (1982a). An empirical Bayesian approach to
   the smooth estimation of unknown functions. MRC
   Technical Report No. 2339.

5. _____ (1982b). Bayes estimation of a multivariable
   density. MRC Technical Report No. 2342.

6. _____ (1982c). Applications of the EM algorithm to
   the estimation of Bayesian hyperparameters. MRC
   Technical Report No. 2344.

Mathematics Research Center and
Department of Statistics
University of Wisconsin-Madison
Madison, WI 53706

Parametric Empirical Bayes Confidence Intervals

Carl N. Morris

I. INTRODUCTION

We consider the problem of estimating $k \geq 3$ unobserved means $\theta_1, \ldots, \theta_k$ of independent normally distributed observed variables $Y = (Y_1, \ldots, Y_k)$

$$Y_i | \theta_i \sim N(\theta_i, V_i) \quad \text{indep.} \tag{1.1}$$

with $V_i = \text{Var}(Y_i | \theta_i)$ assumed known. In practice Y_i might be the sample mean of n_i replications and $V_i = \sigma^2/n_i$.

Empirical Bayes modeling also assumes distributions π for the parameters $\theta = (\theta_1, \ldots, \theta_k)$ exist, with π taken from a known class Π of possible parameter distributions. In this paper we will always let Π be the independent $N(\mu, A)$ distributions on R^k,

$$\theta_i \sim N(\mu, A) \quad \text{indep.,} \quad i=1, \ldots, k. \tag{1.2}$$

Thus each $\pi \in \Pi$ corresponds to one value of (μ, A), $\mu \in R$ and $A \geq 0$ being the prior mean and variance. We also call this a "parametric empirical Bayes" problem because $\pi \in \Pi$ is determined by the parameters (μ, A), and so is a parametric family of distributions.

The usual estimator of θ is the sample mean Y itself, Y being "minimax" for total mean weighted squared error risk

$$E_\theta \sum_1^k (Y_i - \theta_i)^2 W_i = \sum_1^k V_i W_i . \qquad (1.3)$$

This means that no other estimator has smaller risk every-where by any preassigned positive amount. The non-negative constant weights $W_1, \ldots, W_k$ may be chosen arbitrarily in (1.3).

We define the "equal variances case" for this decision problem to mean that the within groups variances $V_i = \mathrm{Var}(Y_i | \theta_i)$ are all equal.

In the equal variances case $V_1 = \cdots = V_k = V$ (say) the statistician clearly should choose $W_1 = W_2 = \cdots = W_k$ in the loss function (1.3). In the "unequal variances case" (meaning $V_i \neq V_j$ for some $i < j$), the choice $W_1, \ldots, W_k$ is ambiguous. Should they be $W_i = 1$ for all i, or $W_i = V_i^{-1}$? Different authors make different choices about this.

Stein (1956, 1961) showed for the equal variances case $V_i \equiv V$ that estimators of the form

$$\hat\theta_i = (1 - \hat B) Y_i + \hat B \mu \qquad (1.4)$$

exist that uniformly dominate $\underline{Y}$ in the sense that if $k \geq 3$

$$E_\theta \sum_1^k (\hat\theta_i - \theta_i)^2 < E_\theta \sum_1^k (Y_i - \theta_i)^2 \qquad (1.5)$$

for all $\underline\theta$. The $\underline\theta$ subscript on E indicates that $\underline\theta$ is fixed in the calculation, expectation being only over the distribution (1.1).

Stein chose

$$\hat B = (k-2) V / \Sigma (Y_i - \mu)^2 \qquad (1.6)$$

with μ a guess at $\bar\theta = \Sigma \theta_i / k$. This shrinking constant, and the estimator (1.4), arise naturally in the empirical Bayes situation, $\hat B$ being an unbiased estimate of $B = V/(V+A)$ under (1.1), (1.2), c.f. Efron and Morris (1973).

Stein's estimator (1.4), (1.6) cannot be used in most practical data analyses, however, usually because the equal variances assumption does not hold. Minimax generalizations of Stein's estimator are possible for the unequal variances situation, e.g. (Bhattacharya (1966), Hudson (1974), Berger (1975)), but the required rule depends on the W_i which are never known in practice, and furthermore, the shrinking patterns of minimax rules do not correspond to exchangeable prior information.

Because of the failure of minimax theory, data analysts usually have chosen to view Stein's estimator in an empirical Bayes context, deriving an appropriate generalization from a parametric empirical Bayes theory, c.f. Carter and Rolph (1974), Efron-Morris (1975), Fay and Herriot (1979), Rubin (1980), Dempster, Rubin, Tsutakawa (1980), Hoadley (1981). The resulting methods may still be "empirical Bayes minimax", but the classical minimax property is sacrificed.

The classical decision theoretic approach also has been unwieldy for providing a confidence set theory associated with Stein's estimator, even in the equal variances case. Some progress has been made concerning confidence ellipsoids by Stein (1962, 1981), Faith (1978), Berger (1980), and Hwang and Casella (1981).

Because classical decision theory has provided a cumbersome and overly restrictive view of shrinking methods, in theory and in application, I prefer to abandon it for these problems in favor of an empirical Bayes theory. Stein's estimator then should be viewed as a centerpiece of (parametric) empirical Bayes theory because it is the rule in the simplest setting (1.1)-(1.2) with equal variances for which mathematical results are most easily derived.

The term "empirical Bayes" is used here to cover a broad class of problems for which a restricted class Π of possible prior distributions is known. Robbins (1955, 1964) adopted this expression to cover a series of problems in which the Bayes risk could be achieved as the number of problems increases without bound. In view of results

established by Efron and Morris for the parametric empirical
Bayes case, i.e. a limited number of estimation problems, it
now makes sense to consider Robbins' work as "nonparametric
empirical Bayes", the results of Efron and Morris as "parametric empirical Bayes", and use the term "empirical Bayes"
to include both viewpoints.

Although it is closely connected with Bayesian ideas,
empirical Bayes inference differs from Bayes because while
a prior distribution is assumed to exist, it is not assumed
known. The global properties established for Stein's estimator, e.g. by Stein (1961) in the classical setting and by
Efron-Morris (1973, 1975) in the empirical Bayes setting,
indicate that Stein's estimator will be successful for
classes of priors like those of (1.2), so the prior need
not be known.

Bayesians might prefer to think of empirical Bayes
inference as robust Bayes, robust to an inappropriate choice
of the prior. This interpretation can be adopted only if
Bayesians are willing to integrate over the sample space,
a practice they traditionally have opposed.

This paper is concerned with existence of "EB confidence
intervals" for the situation (1.1), (1.2) in the equal
variances case, what that means, and how they work on data.
Section 2 presents enough definitions in empirical Bayes
inference to show how EB confidence intervals are defined.
An equal variances data set is analysed in Section 3 using
a proposed shrinking estimator and its confidence interval.
The intervals in this case are 37 percent shorter than standard intervals. The final sections are concerned with deriva-
tions and arguments concerning methods presented in Section
3. The particular problem considered here is very similar
to that in (Morris, 1977), but this paper goes further by
defining empirical Bayes confidence sets and then arguing
that the regions considered have the desired property.
Sections 4 and 5 treat theoretical issues concerning the
derivation and EB confidence properties of the proposed rules.

2. EMPIRICAL BAYES INFERENCE AND CONFIDENCE INTERVALS

Suppose we observe Y with parameter θ and likelihood function $f_\theta(y)$, both y and θ permitted to be multi-dimensional, $Y \varepsilon \mathcal{Y}$ and $\theta \varepsilon \Theta$. Then let Π be a specified class of distributions on Θ, and assume that $\theta \sim \pi$ for some $\pi \varepsilon \Pi$. This is an empirical Bayes (EB) inference problem if an inference is to be made about the realized value of θ and if the operating characteristics of the resulting procedure $t(Y)$ are to be evaluated by averaging over <u>both</u> variables Y and θ.

Thus we observe

$$Y|\theta \sim f_\theta(y) \quad \text{and} \quad \theta \sim \pi, \quad \pi \varepsilon \Pi \tag{2.1}$$

and then use

$$r(\pi,t) = E_\pi E_\theta L(\theta, t(Y)) \tag{2.2}$$

to evaluate the decision procedure t, both Y and θ random. Two special cases are well-known.

Robbins (1956, 1964) used this framework $\theta = (\theta_1, \ldots, \theta_k)$ and $Y = (Y_1, \ldots, Y_k)$ to define decision rules $t_k = t_k(Y_1, \ldots, Y_k)$ converging to the Bayes rule as $k \to \infty$ in certain problems when (Y_i, θ_i) are independent identically distributed (iid), Π being (nearly) the entire class of iid priors on the real line. We will call this an example of "non-parametric empirical Bayes" (NPEB) inference because the prior on one coordinate is unrestricted.

Efron and Morris (1972, 1973, 1975) considered this problem when Π is the subset of Robbins priors that are parametrized by (μ, A) as in (1.2) with $f_\theta(y)$ as in (1.1). The major results there concern estimation with squared error loss, looking for dominance over standard procedures with k fixed. This is an example of "parametric empirical Bayes" (PEB) estimation because Π is parameterized.

Robbins' term "empirical Bayes", which he used for non-parametric empirical Bayes problems, actually fits the

parametric empirical Bayes case too. Hence, the adoption
of the more general meaning for the term "empirical Bayes".
The term also is congenial to Bayesians, e.g. Rubin (1980).

In case Π contains all point priors, i.e. the atom
measures, then we have a classical inference problem and
$r(\pi,t)$ reduces to the ordinary risk

$$r(\pi_\theta,t) = E_\theta L(\theta,t(Y)) \qquad (2.3)$$
$$= \int L(\theta,t(y)) \, f_\theta(y)dy$$

if $\pi = \pi_\theta$ puts all its mass on θ.

At the other extreme, a Bayesian using his unique subjective prior π_0 would consider Π to have only one
element. Then the optimum t is the Bayes rule, choosing
t to minimize

$$E_{\pi_0} L(\theta,t(Y)) \,|\, Y \qquad (2.4)$$

using the distribution of θ given Y and π_0.

An EB confidence set of size $1-\alpha$ is a set $t_\alpha(Y) \subset \Theta$,
Θ the parameter space, such that

$$P_\pi(\theta \,\varepsilon\, t_\alpha(Y)) \geq 1-\alpha, \quad \text{all} \quad \pi \,\varepsilon\, \Pi \qquad (2.5)$$

with (2.5) including expectation over both θ and Y.
Classical confidence sets require

$$P_\theta(\theta \,\varepsilon\, t_\alpha(Y)) \geq 1-\alpha, \quad \text{all} \quad \theta \,\varepsilon\, \Theta \qquad (2.6)$$

choosing Π here to include all point priors, π_θ. Thus
classical confidence sets also are EB confidence sets for
any Π.

In the case of (1.1) $t_\alpha^{(1)}(\underset{\sim}{Y})$ would be a PEB confidence
set for θ_1 wrt Π given by (1.2) if

$$P(\theta_1 \,\varepsilon\, t_\alpha^{(1)}(\underset{\sim}{Y}) \,|\, \mu,A) \geq 1-\alpha, \qquad (2.7)$$

for all $-\infty < \mu < \infty$ and $A \geq 0$.

The objective here is to find intervals for the equal
variances case $V = V_1 = \cdots = V_k$ that satisfy (2.7) for all

μ and $A \geq 0$ and are shorter than the standard intervals $Y_1 \pm z\sqrt{V}$ for θ_1.

3. EB CONFIDENCE INTERVALS FOR 18 BASEBALL BATTERS

Table 1 shows the slightly adjusted* batting averages Y_i of $k = 18$ baseball players in 1970 after each had batted 45 times. These data were analysed earlier, Efron and Morris (1975), and are used again here because a simple example with equal, known variances is needed, with known true values.*

Table 1

THE MAXIMUM LIKELIHOOD ESTIMATES Y_i, EMPIRICAL BAYES ESTIMATES $\hat{\theta}_i$ FOR EACH OF EIGHTEEN BASEBALL PLAYERS

i	Y_i	$\hat{\theta}_i$	s_i
1	0.395	0.308	0.046
2	0.375	0.301	0.044
3	0.355	0.295	0.043
4	0.334	0.288	0.042
5	0.313	0.281	0.041
6	0.313	0.281	0.041
7	0.291	0.274	0.040
8	0.269	0.267	0.040
9	0.247	0.260	0.040
10	0.247	0.260	0.040
11	0.224	0.252	0.040
12	0.224	0.252	0.040
13	0.224	0.252	0.040
14	0.224	0.252	0.040
15	0.224	0.252	0.040
16	0.200	0.244	0.041
17	0.175	0.236	0.043
18	0.148	0.227	0.045
MEAN	0.266	0.266	0.042

*Actually the values Y_i in Table 1 are minor adjustments to the observed averages after 45 appearances given by $Y_i = 0.4841 + 0.0659\sqrt{45}$ arcsin $(2\hat{p}_i-1)$, rounded to three significant figures. The observed average actually is $\hat{p}_i$; for example, $\hat{p}_1 = 18/45 = 0.400$ for player 1 (Roberto Clemente). The arcsin transformation stabilizes variances, as required for assumption (2.1), and the constants 0.4841 and 0.0659 are chosen so that the $\{Y_i\}$ and the $\{\hat{p}_i\}$ have the same mean (0.26567) and standard deviation (0.0659). The same transformation $\theta_1 = 0.4841 + 0.0659\sqrt{45}$ arcsin $(2p_i-1)$ was made to the true values p_i, being the proportion of successes during the remainder of the season for batter i. The names of the players and other information about this problem are contained in Efron-Morris (1975).

Here the true values are the averages during the season's remainder, $\theta_1, \ldots, \theta_k$, which we will view as a random $N(\mu, A)$ sample.[*]

Thus with $V = (.0659)^2$ we assume

$$Y_i | \theta_i \sim N(\theta_i, V), \theta_i | \mu, A \sim N(\mu, A), \quad \text{indep.} \tag{3.1}$$

or equivalently, by Bayes rule, with $B \equiv V/(V+A)$,

$$\theta_i | Y_i, \mu, A \sim N((1-B)Y_i + B\mu, V(1-B)) \quad \text{indep.} \tag{3.2}$$

$$Y_i | \mu, A \sim N(\mu, V+A) \quad \text{indep.} \tag{3.3}$$

Let $\bar{Y} = \Sigma Y_i / k$ and $S = \Sigma (Y_i - \bar{Y})^2$. From standard theory

$$\bar{Y} \sim N(\mu, \frac{1}{k}(V+A)), \quad S \sim (V+A)\chi^2_{k-1}. \tag{3.4}$$

Unbiased estimates of μ, B in (3.2) then are

$$\hat{\mu} = \bar{Y}, \quad \hat{B}_{JS} = 2mV/S, \quad m \equiv (k-3)/2 \tag{3.5}$$

because from (3.3)

$$E\hat{\mu} = \mu, \quad E\hat{B}_{JS} = B. \tag{3.6}$$

Stein's rule is (3.2) with values (3.6) substituted,

$$\hat{\theta}_i = (1-\hat{B}_{JS})Y_i + \hat{B}_{JS}\hat{\mu}. \tag{3.7}$$

Stein notes it is better to replace $\hat{B}_{JS}$ by 1 if $\hat{B}_{JS} > 1$, since $B \leq 1$ always.

The preceding with these data gives $k = 18$, $m = 7.5$, $V = (.0659)^2$, $\hat{\mu} = \bar{Y} = .26567$, $S = 18.93244V$, $\hat{B}_{JS} = .79229$, and so (3.7) gives

$$\hat{\theta}_i = .208Y_i + .2105 \tag{3.8}$$

Bayesian theory may be used to obtain a more complete analysis by estimating B using prior information with μ having a flat prior and A distributed uniform $[0, \infty)$, choices justified in Section 4, giving the Bayes estimator

$$E\theta_i | Y = (1-\hat{B})Y_i + \hat{B}\bar{Y}, \tag{3.9}$$

---

[*]The sampling method used to choose batters after 45 appearances, required to justify the equal variances case, probably acts against this assumption, but not necessarily.

$$\bar{Y} = \sum_i^k Y_i/k, \quad \text{and}$$

$$\hat{B} = \frac{2mV}{S}(1-1/M_m(T)) , \qquad T = S/2V \tag{3.10}$$

and $M_m(T)$ the moment generating function of a Beta $(1,m)$ distribution at T

$$M_m(T) = \int_0^1 \exp[(1-B)T] \, dB^m \tag{3.11}$$

For these data we compute $T = 9.46622$, $M_{7.5}(T) = 6.77428$, $\hat{B} = \hat{B}_{JS}(.85238) = 0.67534$. Thus (3.9) gives the estimate

$$\hat{\theta}_i = .325Y_i + .1794 \tag{3.12}$$

Assuming the prior distribution for μ, A, we also use for the variance of the estimator

$$s_i^2 \equiv \text{Var}(\theta_i | \underset{\sim}{Y}) = V(1 - \frac{k-1}{k}\hat{B}) + v(Y_i - \bar{Y})^2 \tag{3.13}$$

with

$$v \equiv \text{Var}(B|S) = -\frac{\partial}{\partial T}\hat{B}$$

$$= \frac{1}{m}\hat{B}^2 - (\hat{B}_{JS} - \hat{B})(1 - \frac{m+1}{m}\hat{B}) \tag{3.14}$$

$$= .0608 - .0274 = (.1827)^2 .$$

Thus

$$s_i = .0397[1 + 21.22(Y_i - .266)^2]^{1/2} \tag{3.15}$$

$$R^* = \frac{1}{k}\sum_1^k s_i^2/V = .397 = (.630)^2 , \tag{3.16}$$

Finally, the unbiased estimate $\hat{R}$ of the risk $E_\theta \sum_1^k (\hat{\theta}_i - \theta_i)^2/kV$ for this estimate is

$$\hat{R} = R^* - (2m - \hat{B}S/V)/k = .274 , \tag{3.17}$$

suggesting this estimator is $3.6 = 1/.274$ times as efficient in terms of mean squared error as the sample means, over 18 coordinates.

The values Y_i, $\hat{\theta}_i$, s_i are recorded in Table 1.

It is not true that θ_i given $\underset{\sim}{Y}$ has exactly a normal distribution, but it is true that θ_i given both (Y,B) from (3.2) is normal. The normality of $\underset{\sim}{Y}$ from (3.3), considered as an added source of variation in the estimation rule, further suggests that interval estimates of θ_i might be based on normal assumptions, and we will do that here. Other authors, in commenting on this rule as presented in (Morris, 1977), have noted that this approximation is imperfect (Berger, 1980), and have suggested improvements by use of Edgeworth expansions (Van der Merwe, et. al., 1981). Such modifications complicate our main task and must assume the prior distribution chosen for A is valid. While those ideas are important, they will be ignored here.

We will make and support the claim that

$$P(\hat{\theta}_i - zs_i \le \theta_i \le \hat{\theta}_i + zs_i) \ge 2\Phi(z) - 1. \qquad (3.18)$$

Thus, for example, $\theta_i \pm 1.96s_i$ can be considered as a .95 EB confidence interval, $\Phi(z)$ being the $N(0,1)$ distribution function. Evidence for this claim is offered in Section 5. Statement (3.18) is true if it holds for all (μ,A) in the sense of (2.7). The derivation of $\hat{\theta}_i,s_i$ has been based on the flat prior for μ, and independently on $A \sim \text{Uniform}[0,\infty)$. Thus (3.18) holding for all μ,A is a strong conclusion in need of independent verification.

In view of (3.18) we can make the following EB confidence statement:

Assuming the players are sampled from a population with normally distributed batting abilities, then we are 95% EB confident that

$$\hat{\theta}_1 - 1.96s_1 \le \theta_1 \le \hat{\theta}_1 + 1.96s_i \qquad (3.19)$$

i.e. that $.218 \le \theta_1 \le .398$.

Similar statements may be made about θ_2, etc.

Note that player 1 actually batted .395. The information carried by the other 17 players indicate that player 1 partly was lucky to do so well, and should not expect to continue at that pace in the long run.

Formally, these are not probability statements, because the distributions of both θ and Y are used in derive (3.18), and an interval for θ_i would have to hold for fixed Y_i to be a probability statement. Thus the term "confidence" is used.

The intervals (3.18) are not confidence intervals in the usual sense. For example if $\theta_2 = \theta_3 = \cdots = \theta_k$ and $\theta_1 = \theta_2 + \sqrt{kV}$ then we would observe $s_1^2 \doteq V$ and $E_\theta (\hat{\theta}_1 - \theta_1)^2 \doteq Vk/4$. Thus the claim that $(\hat{\theta}_1 - \theta_1)^2 \leq (1.96)^2 s_1^2$ with high probability is badly violated for large k.

We do not believe in the prior chosen for (μ, A). Rather, it was used to derive a rule with good properties. This prior was chosen for five reasons:

(1) Stein's estimator is a formal Bayes rule assuming μ is flat and $A \sim \text{Uniform}[-V, \infty)$. Removing the possibility that the variance $A < 0$ is an obvious improvement, hence $A \sim \text{Uniform}[0, \infty)$.

(2) The risk R^* in (3.16) and $\hat{R}$ in (3.17) satisfy $R^* > \hat{R}$ for all samples. Thus, in the mean square sense, s_i is too wide whatever the true θ_i, and thus conservative:

$$kE_\theta \hat{R} = \sum_1^k E_\theta (\hat{\theta}_i - \theta_i)^2 < E_\theta \sum_1^k s_i^2 = kER^*_\theta \qquad (3.20)$$

for all θ.

(3) Similarly, $\hat{R}$ and $R^* < 1$ for all samples. Thus the estimator is "minimax" in the usual sense that

$$\frac{1}{k} E_\theta \sum_1^k (\hat{\theta}_i - \theta_i)^2 = E_\theta \hat{R} < 1. \qquad (3.21)$$

(4) It also follows that the estimator is "EB minimax" for each coordinate, e.g. for all i

$$E(\hat{\theta}_i - \theta_i)^2 = E\hat{R} < 1 \qquad (3.22)$$

with expectation over both $\underset{\sim}{Y}$ and $\underset{\sim}{\theta}$ according to (3.1), (3.2).

(5) The prior does not depend on the variance V. Some proposed priors do, e.g. Strawderman (1971), making them awkward to use in the unequal variances case. This prior has been considered previously by several authors: Stein (1962), Baranchik (1964), as a priori on $\|\underset{\sim}{\theta}\|^2$ and by Leonard (1974).

Note that the precision indicated by s_i is better for components Y_i near $\bar{Y}$. This is in agreement with the familiar regression result that prediction errors are smaller near the mean of the explanatory variables. Thus the smallest s_i is .040, but batter number 1 has $s_1 = .046$, see Table 1.

The typical value of s_i may be taken to be the root mean square $\sqrt{VR^*} = .0415$, and since $.0415/\sqrt{V} = .630$, the intervals of Table 1 are 37 percent shorter while covering the true values with at least the same confidence as the usual intervals.

How can this be? Table 2 has the true values $\theta_1, \ldots, \theta_{18}$ and Figure 1 plots them against $Y_1, \ldots, Y_{18}$. The slope of the true θ_i versus Y_i is much less than unity in Figure 1 because of regression to the mean at both ends of the chart. We have $C = \sum_1^k \theta_i (Y_i - \bar{Y}) = .01525$ and so the best linear predictor in this case would be the regression line

$$\hat{\theta}_i = \bar{Y} + \frac{C}{S}(Y_i - \bar{Y})$$

$$= .266 + (1 - .815)(Y_i - .266) \qquad (3.23)$$

This cannot be computed in practice, but it shows the optimal shrinkage coefficient would be .815, even more than the .675

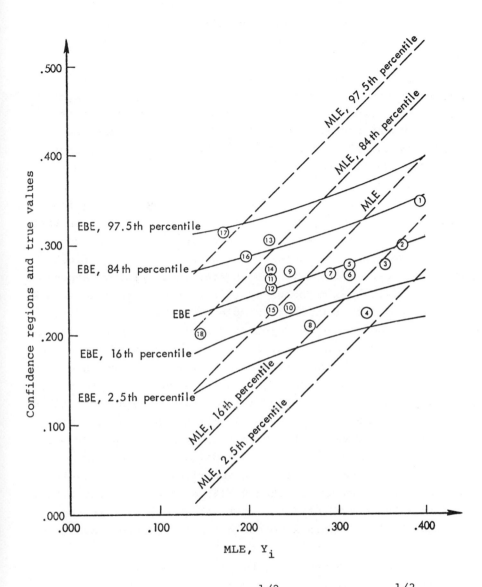

Figure 1. MLE Y_i, MLE $\pm V^{1/2}$, and MLE $\pm 1.96V^{1/2}$
(dashed lines) and $\hat{\theta}_i = EBE$, EBE $\pm s_i$, and EBE $\pm 1.96s_i$
(solid curves). Eighteen players plotted at (Y_i, θ_i)
using data of Tables 1,2.

estimated. The optimal slope is .185, or 11 degrees, far
from 45 degrees for the maximum likelihood estimator Y_i
(MLE).

Note that 13/18 of the true values in Figure 1 are
within the one standard deviation bound for the EB estimator,
and 12/18 for the MLE, versus 12.3 expected (68.3 percent).
The 95 percent bounds contain all 18 true values for the EB
estimator, and 17/18 for the MLE. Thus the empirical Bayes
intervals are simultaneously more accurate and 37 percent
shorter in this case.

An extremely interesting point raised by Figure 1 is
that if the 95 percent confidence region is used to test that
the true value of a player is a specified value, then con-
flicting results occur between the classical and empirical
Bayes methods. Because it has shorter intervals, we expect
the empirical Bayes methodology to reject certain true values
when the MLE does not. For example, from Figure 1, a 0.500
season average cannot be rejected for player number 1 according to classical theory, but is out of the question from the
empirical Bayes standpoint. No one ever has approached such
a value for a full season. The surprising fact is that the
empirical Bayes method includes two small regions excluded by
the classical methodology. To illustrate this, a true value
of 0.318 is rejected at the 95 percent level for player number 17 (Thurman Munson) by the classical test, but is not
rejected at the same level using empirical Bayes intervals
in Table 1. It turns out that 0.318 was Munson's true value.
(And in 1976 he was voted the most valuable player in the
American League!) This hypothesis testing problem obviously
is a topic worth of further research.

Table 2 also shows that the $(\hat{\theta}_i - \theta_i)/s_i$ values are
approximately $N(0,1)$ in this example. They are correlated
with the Y_i values, however, because $\hat{B}$ is smaller than
the optimal value.

A simulation was used to determine that the intervals
$\hat{\theta}_i \pm s_i$ and $\hat{\theta}_i \pm 1.96 s_i$ contain the true values θ_i in at

Table 2

TRUE VALUES, RELATIVE ERRORS, AND LOSSES FOR
EMPIRICAL BAYES ESTIMATES AND
MAXIMUM LIKELIHOOD ESTIMATES

(1)	(2)	(3)	(4)	(5)	(6)	(7)
	TRUE VALUE	EBE RELATIVE ERROR	MLE RELATIVE ERROR	EBE LOSS	MLE LOSS	
i	θ_i	$\dfrac{\hat{\theta}_i - \theta_i}{s_i}$	$\dfrac{Y_i - \theta_i}{\sqrt{V}}$	$\dfrac{(\hat{\theta}_i - \theta_i)^2}{V}$	$\dfrac{(Y_i - \theta_i)^2}{V}$	N^*
1	0.346	-0.831	0.744	0.339	0.553	367
2	0.300	0.026	1.138	0.000	1.295	426
3	0.279	0.365	1.153	0.057	1.330	521
4	0.223	1.560	1.684	0.968	2.837	275
5	0.276	0.124	0.561	0.006	0.315	418
6	0.273	0.198	0.607	0.015	0.368	466
7	0.266	0.198	0.379	0.014	0.144	586
8	0.211	1.406	0.880	0.716	0.775	138
9	0.271	-0.286	-0.364	0.030	0.133	510
10	0.232	0.694	0.228	0.175	0.052	200
11	0.266	-0.343	-0.637	0.044	0.406	277
12	0.258	-0.145	-0.516	0.008	0.266	270
13	0.306	-1.334	-1.244	0.668	1.548	435
14	0.267	-0.368	-0.653	0.051	0.426	538
15	0.228	0.598	-0.061	0.134	0.004	186
16	0.288	-1.054	-1.335	0.439	1.783	558
17	0.318	-1.903	-2.170	1.540	4.709	408
18	0.200	0.609	-0.789	0.174	0.623	70
MEAN	0.267	-0.027	-0.022	0.299	0.976	369
STDEV	0.037	0.862	0.988	0.412	1.157	150

*N is number of at bats upon which the "true value" θ_i is computed.

least 68 percent and 95 percent of the cases, using the
$\{\theta_i\}$ from Table 2 each time, but generating new values
$Y_1, \ldots, Y_k$ each time according to (3.1) Based on 1800 experiences, the actual coverage probabilities were 74 percent and
97.3 percent. The average $\hat{B}$ was 0.61 and intervals were
35 percent shorter than the MLE. Thus, the data considered
are typical of what one expects for these true values.

The reader is cautioned, however, that these good re-
sults cannot be expected if the $(\theta_1, \ldots, \theta_k)$ do not follow
an exchangeable distribution, as assumed in (3.1). Empirical
Bayes estimators, or Stein's estimator, can lead to misesti-
mation of components that the statistician or his clients
care about when exchangeability in the prior distribution is
implausible.

4. DERIVATION OF RULES CONSIDERED

We seek to derive the formal Bayes estimator of θ_i and
its variance under assumptions (3.1), (3.2) and with prior
probability element

$$d\mu \; dA \quad -\infty < \mu < \infty \; , \quad A \geq 0 \tag{4.1}$$

on (μ, A). First, starting with (3.2) and keeping A fixed,
$\bar{Y}$ is a sufficient statistic for μ from (3.3) with
$\bar{Y} - \mu \sim N(0, (V+A)/k)$. Thus, in view of the flat prior on μ,
a standard calculation shows

$$\mu | \bar{Y}, A \sim N(\bar{Y}, (V+A)/k). \tag{4.2}$$

It follows from (3.2) that

$$\theta_i | \underline{Y}, A \sim N((1-B)Y_i + B\bar{Y}, V(1 - \frac{k-1}{k} B)) \tag{4.3}$$

because $\text{Var}(\theta_i | \underline{Y}, A) = V(1-B) + \text{Var}(B\mu | Y, A) = V(1-B) + B^2(V+A)/k$.

A sufficient statistic for the unknown B in (4.3) is
$T \equiv S/2V \sim B^{-1}\text{Gam}(m+1, 1)$ with $m \equiv (k-3)/2$. Now since
$B = V/(V+A)$, $dA = -VdB/B^2$, $0 < B \leq 1$. Thus

$$\hat{B} = EB|Y = EB|S$$

$$= \frac{\int_0^1 B\,B^{m-1}\exp(-TB)\,dB}{\int_0^1 B^{m-1}\exp(-TB)\,dB} \tag{4.4}$$

$$= 1 - \frac{d}{dT}\log M_m(T) . \tag{4.5}$$

with $M_m(T)$ as in (3.11).

If we instead had $V+A \sim \text{Unif}[0,\infty)$ then the upper end point of each integral in (4.4) would be ∞, not 1 and $EB|T = m/T$ would follow from (4.4) and $M_m(T) = \Gamma(m+1)T^{-m}$ $\exp(T)$ in this case. Thus the prior (4.6) yields Stein's estimator.

But for the prior $A \sim \text{Unif}[0,\infty)$, which is similar, but more reasonable because it forces $A \geq 0$, integrating the numerator of (4.4) once by parts, using $u = B^m$, $dv = \exp(-TB)\,dB$ yields

$$-\exp(-T)/T + \frac{m}{T}\int_0^1 B^{m-1}\exp(-BT)\,dB \tag{4.6}$$

and hence (4.4) simplifies to

$$EB|S = \hat{B} = \frac{m}{T}\left(1 - \frac{1}{M_m(T)}\right) \tag{4.7}$$

with $M_m(T)$ defined by (3.11).

Furthermore,

$$v \equiv \text{Var}(B|T) = -\frac{d\hat{B}}{dT}$$

$$= \frac{1}{T}\hat{B} - \frac{m}{T}\frac{M'_m(T)}{M^2_m(T)}$$

$$= \frac{1}{T}\hat{B} - \frac{m}{T}\frac{1}{M_m(T)}(1-\hat{B})$$

$$= \frac{\hat{B}}{T} - \frac{1-\hat{B}}{T}(m-T\hat{B})$$

$$= \hat{B}^2/m - (\hat{B}_{JS}-\hat{B})(1-\frac{m+1}{m}\hat{B}) \tag{4.8}$$

$$\leq \hat{B}^2/m \tag{4.9}$$

This follows because $M_m'(\dot{T})/M_m(t) = \frac{d}{dt} \log M_m(T) = 1-\hat{B}$,   and

$\hat{B}_{JS} = m/T$ in (4.8).

It is easy to see that $\hat{B}$ is decreasing in T and that $\lim \hat{B} = m/(m+1)$ as $T \to 0$. Thus the inequality (4.9) is justified, and it is sharp as $T \to 0$ and $T \to \infty$.

It follows from (4.3) that

$$\hat{\theta}_i \equiv E\theta_i | \underset{\sim}{Y} = (1-\hat{B})Y_i + \hat{B}\overline{Y} \tag{4.10}$$

$$s_i^2 \equiv \text{Var}(\theta_i | Y) = V(1 - \frac{k-1}{k} \hat{B}) + v(Y_i - \overline{Y})^2 \tag{4.11}$$

with $\hat{B}$ and v in (4.7) and (4.8).

Had μ been known, not estimated, the previous results would change by letting $S = \Sigma(Y_i-\mu)^2$, $m = (k-2)/2$, $\hat{B}$, v, and $\hat{\theta}_i$ still be given by (4.7), (4.8), and (4.10), and

$$\text{Var}(\theta_i | Y) = V(1-\hat{B}) + v(Y_i-\mu)^2 \tag{4.12}$$

Had Stein's estimator been considered, with the prior (4.6) $A \sim \text{Unif}[-V, \infty)$ then (4.10) and (4.11) or (4.12) hold, but $\hat{B} = m/T = \hat{B}_{JS}$ and $v = \hat{B}^2/m$ from (4.8).

The unbiased estimate of component risk of any estimator of θ_i having the form $\hat{\theta}_i = \overline{Y} + (1-B(T))(Y_i-\overline{Y})$ is, denoting $B'(T) = dB(T)/dT$:

$$\hat{R}_i = 1 - 2\frac{k-1}{k}B(T) + \frac{(Y_i-\overline{Y})^2}{V}[B^2(T) - 2B'(T)], \tag{4.13}$$

i.e.,

$$E_\theta \hat{R}_i = E_\theta(\hat{\theta}_i - \theta_i)^2/V \tag{4.14}$$

which follows from Stein (1973) or Efron-Morris (1976). Thus, if $\hat{\theta}_i$ is a Bayes estimator, $-B'(T) = v = \text{Var}(B|T)$.

$$\hat{R}_i = 1 - 2\frac{k-1}{k}B(T) + \frac{(Y_i-\overline{Y})^2}{V}[B^2(T) + 2v] \tag{4.15}$$

$$\hat{R} = \frac{1}{k}\Sigma\hat{R}_i = 1 - 2\frac{k-1}{k}B(T) + \frac{2T}{k}[B^2(T) + 2v] \tag{4.16}$$

and

$$R^* \equiv \frac{1}{kV} \sum_1 s_i^2 = 1 - \frac{k-1}{k} B + \frac{2vT}{k} \qquad (4.17)$$

$$R^* - \hat{R} = \frac{2m}{kM_m(T)} = \frac{2}{k} (m - T\hat{B}) > 0. \qquad (4.18)$$

In case of the Stein prior (4.6), these expressions hold, formally putting $M_m(T) = \infty$. Thus

$$R^* = \hat{R} = 1 - \frac{2m}{k} \hat{B} . \qquad (4.19)$$

5. <u>EB CONFIDENCE INTERVALS</u>

For simplicity, assume μ is known, and hence zero. This slightly modifies the preceding rule used to analyse the baseball data. Taking $V = 1$, s_i is given by (4.12) as

$$s_i^2 = 1 - \hat{B} + vY_i^2 \qquad (5.1)$$

$$\hat{\theta}_i = (1 - \hat{B})Y_i \qquad (5.2)$$

with $\hat{B} = \frac{m}{T}(1 - 1/M_m(T))$, $m = (k-2)/2$. Given $A \geq 0$ we are to show under (3.2) and (3.3)

$$\theta_i | Y_i \sim N((1-B)Y_i, 1-B) \qquad (5.3)$$

with $B = 1/(1+A)$ and

$$Y_i \sim N(0, 1+A) \quad \text{indep.} \qquad (5.4)$$

that

$$P(\hat{\theta}_1 - zs_1 \leq \theta_1 \leq \hat{\theta}_1 + zs_1)$$

$$= E\{\Phi(W + zQ) - \Phi(W - zQ)\} \qquad (5.5)$$

$$\geq \Phi(z) - \Phi(-z) \equiv 1 - \alpha(z), \qquad (5.6)$$

where $W = (B - \hat{B})Y_1 / \sqrt{1-B}$, $Q^2 = 1 + \dfrac{B - \hat{B} + vY_1^2}{1-B}$ are random variables depending on Y_1 and $S = \Sigma Y_i^2$ only, with distributions given by (5.4).

Proposition 1.

The inequality linking (5.5) and (5.6) is at least approximately true, and thus $(\hat{\theta}_i - zs_i, \hat{\theta}_i + zs_i)$ is an EB confidence interval for each $1 \le i \le k$ under assumptions (3.2), (3.3) and $k \ge 3$, with confidence level $1 - \alpha(z)$ $\equiv \Phi(z) - \Phi(-z)$, Φ the $N(0,1)$ distribution function.

We have no formal proof of Proposition 1, i.e. of the inequality relating (5.5) and (5.6). The following evidence is available.

Let $\pi(k,B,z)$ be the left hand side of (5.5), and $1 - \alpha(z)$ denote (5.6).

(1) $\lim_{k \to \infty} \pi(k,B,z) = 1 - \alpha(z)$ for all (B,z).

This means (5.5) forms EB confidence intervals, in the asymptotic sense of Robbin's empirical Bayes,

(2) $\lim_{B \to 0} \pi(k,B,z) = 1 - \alpha(z)$

for all z, $k \ge 3$.

(3)   For all B, k

$$E_B(\hat{\theta}_i - \theta_i)^2 \le E_B s_i^2 . \tag{5.7}$$

Thus, in expectation, s_i^2 is large enough for (5.5), (5.6) to hold.

(4)   An approximation to (5.5) that is valid for W near 0 and Q near 1 is given by

$$\Phi(W + zQ) - \Phi(W - zQ)$$
$$\doteq 1 - \alpha(z) + z\phi(z)(Q^2 - W^2 - 1) . \tag{5.8}$$

Now

$$(1 - B)(Q^2 - W^2 - 1) = B - \hat{B} + (v - (B-\hat{B})^2)Y_1^2 , \tag{5.9}$$

and it can be shown that

$$E_B(Q^2 - W^2 - 1) = E_B(R^* - \hat{R})/(1 - B) \tag{5.10}$$

$$\ge 0$$

because $R^* > \hat{R}$ for every k. Thus the expectation of the right side of (5.8) exceeds $1 - \alpha(z)$ for all B, $k \geq z$. The expectation of the left side does too if B is near enough to zero or k large enough so that W^2 and $Q^2 - 1$ are likely to be near zero. The preceding results will be proved in another paper on this subject.

(5) Figures 2 and 3 offer other evidence. They show for $k = 8$ and $B = 0.5$ (Figure 2) and $B = 0.8$ (Figure 3) that the one-tailed complement of (5.5), $P(\hat{\theta}_1 - \theta_1 > zs_1)$ $= E\Phi(W - zQ)$ is generally less in a simulation than $.5\alpha(z)$ for $z = 0.675, 1.282, 1.645, 1.960$ $(.5\alpha(z) = .25, .10, .05, .025)$. Each point in the Figures is actually an average of $2k = 16$ values of $\Phi(W - zQ)$ over probability samples of (Y_1, S) and therefore of (W, Q). These are less than $.5\alpha(z)$ when S is moderately near its expectation $(ES = k/B = 16$ in Figure 2, $= 10$ in Figure 3) but rise slightly above $.5\alpha(z)$ when S is small or large in Figure 2. All sampled values fall under $.5\alpha(z)$ in Figure 3, the EB confidence intervals as constructed being more conservative when B is near 1.

The cases with B nearer zero have been checked by simulation when $k = 8$ to be well behaved, as result (2) suggests. Larger values of k give better results than presented here for $k = 8$, as result (1) suggests.

We conclude that Proposition 1 holds.

6. CONCLUSION

A new notion of confidence sets, "EB and PEB confidence sets", appears for the first time here. The baseball example and the rule proposed for that equal variance normal case shows that such sets can be considerably shorter than the standard frequentist intervals, while having higher than nominally claimed confidence.

This rule does not have the frequentist confidence property. It has a stronger property than Bayesian theory claims because the confidence property does not depend on

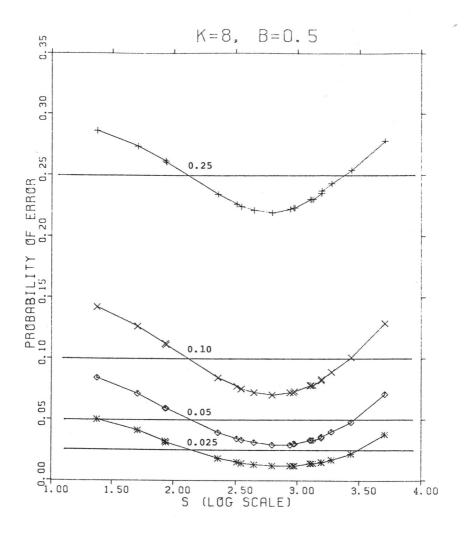

Figure 2. Error probabilities for EB confidence
interval (5.5) with z = .675, 1.282, 1.645, 1.960
giving nominal probabilities of error: α(z)/2 = .25,
.10, .05, .025. See text for fuller explanation.

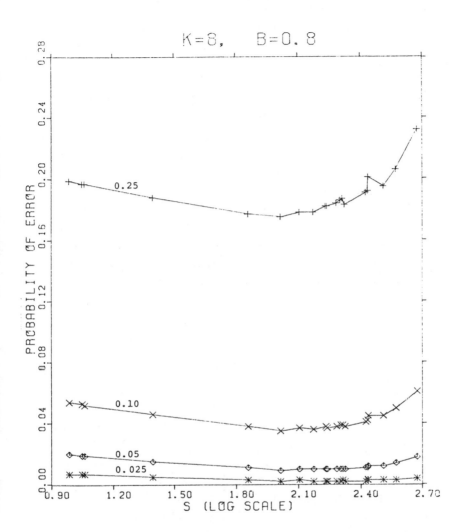

Figure 3. Error probabilities for EB confidence interval (5.5) with z = .675, 1.282, 1.645, 1.960 giving nominal probabilities of error: $\alpha(z)/2 = .25$, .10, .05, .025. See text for fuller explanation.

correctness of a particular prior distribution. Rather, the
confidence property is obtained by integrating over the
sample space so that coverage holds for a class of prior
distributions.

   The statistician using such rules should be warned that
he cannot expect gains over the standard frequentist methods
without exchangeability in the prior distributions. Hence,
the improvement is gained in exchange for increased knowledge.
In some cases, however, this exchangeability assumption
seems warranted, for example, when the individuals being
analysed are a sample from a population. Then empirical
Bayes methods in general, and parametric empirical Bayes
methods in particular, provide a way to utilize this additional information by obtaining more precise estimates and
estimating their precision. A simple case is considered
here; clearly much more remains to be done to make the results
useful in the more complicated settings one ordinarily
encounters in practice.

REFERENCES

Baranchik, A.J., "Multiple Regression and Estimation of the
     Mean of a Multivariate Normal Distribution," Technical
     Report No. 51, Stanford University, Department of
     Statistics, 1964.

Berger, J. O., "Admissible Minimax Estimation of a Multi-
     variate Normal Mean with Arbitrary Quadratic Loss,"
     Ann. Statist., Vol. 4, No. 1, 1976, pp. 223-226.

Berger, J. O., "A Robust Generalized Bayes Estimator and
     Confidence Region for a Multivariate Normal Mean,"
     Ann. Statist., Vol. 8, 1980, pp. 716-761.

Bhattacharya, P. K., "Estimating the Mean of a Multivariate
     Normal Population with General Quadratic Loss Function,"
     Ann. Math. Statist., Vol. 37, 1966, pp. 1819-1824.

Carter, G. and J. Rolph, "Empirical Bayes Methods Applied
     to Estimating Fire Alarm Probabilities," J. Amer.
     Statist. Assoc., Vol. 69 (1974), pp. 880-885.

Dempster, A. P., D. B. Rubin, and R. K. Tsutakawa, "Estimation
     in Covariance Components Models," J. Amer. Statist.
     Assoc., Vol. 76, No. 374 (1981), pp. 341-353.

Efron, B., and C. Morris, "Limiting the Risk of Bayes and Empirical Bayes-Estimators--Part II: The Empirical Bayes Case," J. Amer. Statist. Assoc., Vol. 67, No. 337 (March 1972), pp. 130-139.

Efron, B., and C. Morris, "Stein's Estimation Rule and Its Competitors--An Empirical Bayes Approach," J. Amer. Statist. Assoc., Vol. 68, No. 341 (March 1973), pp. 117-130.

Efron, B., and C. Morris, "Data Analysis Using Stein's Estimator and Its Generalizations," J. Amer. Statist. Assoc., Vol. 70, No. 350, (June 1975), pp. 311-319.

Efron, B., and C. Morris, "Families of Minimax Estimators of the Mean of a Multivariate Normal Distribution," Ann. Statist., Vol. 4, No. 1 (1976), pp. 11-21.

Faith, R. E., "Minimax Bayes Estimators of a Multivariate Normal Mean," J. Multivariate Anal., Vol. 8 (1978), pp. 372-379.

Fay, R. E. III, and R. A. Herriot, "Estimates of Income for Small Places: An Application of James-Stein Procedures to Census Data," J. Amer. Statist. Assoc., Vol. 74 (1979), pp. 269-277.

Hoadley, B., "Quality Management Plan (QMP)," Bell System Technical Journal, Vol. 60 (1981), pp. 215-273.

Hudson, H. M., "Empirical Bayes Estimation," Stanford University, Department of Statistics, Technical Report No. 58, May 31, 1974.

Hwang, J. T. and G. Casella, "Minimax Confidence Sets for the Mean of a Multivariate Normal Distribution," Department of Mathematics, Cornell University, August 1981, Ann. Statist., to appear.

James, W., and C. Stein, "Estimation with Quadratic Loss," Proceedings of the Fourth Berkeley Symposium on Mathematical Statistics and Probability, Vol. 1, University of California Press, Berkeley (1961), pp. 361-379.

Leonard, T., "A Bayesian Method for the Simultaneous Estimation of Several Parameters," University of Warwick, Statistics Department, Coventry, England, April 1974.

Morris, C., "Interval Estimation for Empirical Bayes General-
    izations of Stein's Estimator," Proc. of Twenty-second
    Conf. on the Design of Experiments in Army Research
    Development and Testing, (1977), ARO Report 77-2.

Robbins, H., "The Empirical Bayes Approach to Statistical
    Decision Problems," Ann. Math. Statist., Vol. 35, No. 1,
    (March 1964), pp. 1-20.

Robbins, H., "An Empirical Bayes Approach to Statistics,"
    Proc. Third Berkeley Symposium Math. Statist. Prob.,
    Vol. 1 (1955), pp. 157-164.

Rubin, D., "Using Empirical Bayes Techniques in the Law
    School Validity Studies," Jrnl. Amer. Statist. Assoc.,
    Vol. 75, No. 372 (1981), pp. 801-827, with discussion.

Stein, C., "Confidence Sets for the Mean of a Multivariate
    Normal Distribution," J. Royal Statist. Soc., Series B,
    Vol. 24, No. 2 (1962), pp. 265-296.

Stein, C., "Estimation of the Mean of a Multivariate Normal
    Distribution," Proceedings of the Prague Symposium on
    Asymptotic Statistics," September 3-6, 1973, pp. 345-
    381.

Stein, C. M., "Estimation of the Mean of a Multivariate
    Normal Distribution," Annals Statist., Vol. 9, No. 6,
    (November 1981), pp. 1135-1151.

Strawderman, W. E., "Proper Bayes Minimax Estimators of the
    Multivariate Normal Mean," Ann. Math. Statist., Vol. 42,
    No. 1 (1971), pp. 385-388.

Van der Merwe, A. J., P. C. N. Groenewald, D. G. Nel, C. A.
    Van der Merwe, "Confidence Intervals for a Multivariate
    Normal Mean in the Case of Empirical Bayes Estimators
    Using Pearson Curves and Normal Approximations," Univer-
    sity Orange Free State, Department of Math. Statistics,
    Republic of South Africa, Technical Report No. 70
    (1981).

This research was supported by the National Science Foundation
under Grant number MCS-8104-250.

                              Department of Mathematics
                              University of Texas at Austin
                              Austin, TX 78712

An Apology for Ecumenism in Statistics

G. E. P. Box

Perhaps I should begin with an apology for my title. These days the statistician is often asked such questions as "Are you a Bayesian?" "Are you a frequentist?" "Are you a data analyst?" "Are you a designer of experiments?" I will argue that the appropriate answer to <u>all</u> these questions can be (and preferably should be) "yes", and that we can see why this is so if we consider the scientific context of what statisticians do.

For many years Statistics has seemed to be in a rather turbulent state and the air has been full of argument and controversy. The relative virtue of alternative methods of inference and, in particular, of Bayes' and Sampling (frequentist) inference has been hotly debated. Recently Data Analysis has rightly received much heavier emphasis, but its more avid proponents have sometimes seemed to suggest that all else is worthless. Furthermore while biased estimators, in particular shrinkage and ridge estimators, which have been advocated to replace the more standard varieties are clearly sensible in appropriate contexts their frequentist justification which ignores context seems unconvincing. Parallel criticism may be made of ad hoc robust procedures the proliferation of which has worried some dissidents who have argued for example that mechanical downweighting of peculiar observations may divert attention from important clues to new discovery.

    Insofar as these debates lead us to progressive change
in our ideas they are healthy and productive, but insofar as
they encourage polarization they may not be. One remembers
with some misgivings Saxe's poem about the six blind men of
Hindustan investigating an elephant. It will be recollected
that one, feeling only the elephant's trunk, thought it like
a snake, another, touching its ear, thought it must be a
fan, etc. The poem ends:

>   And so these men of Hindustan
>   Disputed loud and long,
>   Each in his opinion
>   Exceeding stiff and strong,
>   Though each was partly in the right,
>   And all were in the wrong.

    Some of the difficulties arise from the need to
simplify. But simplification included merely to produce
satisfying mathematics or to reduce problems to convenient
small sized pieces can produce misleading conclusions.
Simplification which retains the essential scientific
essence of the problem is most likely to lead to useful
answers but this requires understanding of, and interest in,
scientific context.

1.   SOME QUESTIONABLE SIMPLIFICATIONS.

    (a) It has been argued that Bayes' theorem uniquely
solves all problems of inference. However only part of the
inferential exercises in which the statistical scientist is
ordinarily engaged seem to conveniently fit the Bayesian
mold. In particular diagnostic checks of goodness of fit
involving various analyses of residuals seem to require
other justification. In fact I believe (Box [1980]) that
the process of scientific investigation involves not one but
two kinds of inference: estimation and criticism, used
iteratively and in alternation. Bayes completely solves the
problem of estimation and can also be helpful at the
criticism stage in judging the relative plausibility of two
or more models. However because of its necessarily
conditional nature, it cannot deal with the most essential
part of inferential criticism which requires a sampling
(frequentist) justification.

(b) Fisher [1956] believed that the Neyman-Pearson
theory for testing statistical hypotheses, while providing a
model for industrial quality control and sampling
inspection, did not of itself provide an appropriate basis
for the conduct of scientific research. This can be
regarded as the complement to the objection raised in (a),
for statistical quality control and inspection are methods
of inferential criticism supplying a continuous check on the
adequacy of fit of the model for the properly operating
process. I would regard Fisher's comment as meaning that
the Neyman-Pearson theory was irrelevant to problems of
estimation. Certainly there is evidence in the social
sciences that excessive reliance upon this theory alone,
encouraged by the mistaken prejudices of referees and
editors, has led to harmful distortion of the conduct of
scientific investigation in these fields.

(c) In some important contexts the scientific
relevance of alphabetic optimality criteria (A,E,D,G etc.)
in the choice of experimental designs has been questioned
(see discussion of Kiefer [1959], also Box [1982]). Here
again there is danger of deleterious feedback since users of
statistical design, perhaps dazzled by impressive but poorly
comprehended mathematics, may fail to realize the naive
framework within which the optimality occurs.

(d) Even Data Analysis, excellent in itself, presents
some dangers. It is a major step forward that in these days
students of statistics are required more and more to work
on real data. Indeed suitable "data sets" have been set
aside for their study. But this too can produce misunderstanding. For instance, some examples have become notorious
and have been analyzed by a plethora of experts; one
finds three outliers, another claims that a transformation
is needed and then only one outlier occurs, and so on.
Too much exposure to this sort of thing can again lead to
the mistaken idea that this represents the real context
of scientific investigation. The statistician in his
proper role as a member of a scientific team should
certainly make such analyses, but realistically he would
then discuss them with his scientific colleagues and

present, when appropriate, not one, but alternative
plausible possibilities. He need not, and usually should
not, choose among them. Rather he should make sure that
these possibilities were considered when he and his
scientific colleagues planned the next stage of the
investigation. Together they would choose the next design
so that among other things it could resolve current
uncertainties judged to be important. In particular the
possible meaning and importance of discrepant values would
then be discussed as well as the meaning of analyses which
downweighted or excluded them.

The most dangerous and misleading of the unstated
assumptions suggested to some extent by all these
simplifications concerns the implied <u>static</u> nature of the
process of investigation: <u>A</u> Bayesian analysis is made; <u>a</u>
hypothesis is tested; <u>one</u> model is considered; <u>a single</u>
design is run; <u>a single</u> set of data is examined and
reexamined[1].

I believe that the object of statistical theory should
be to explain, at least approximately, what good scientists
do and to help them do it better. It seems necessary
therefore to examine at least briefly the nature of the
scientific process itself.

2.   SCIENTIFIC METHOD AND THE HUMAN BRAIN.

Scientific method is a formalization of the everyday
process of finding things out. For thousands of years,
things were found out largely as a result of chance
occurrences. For a new "natural law" to be discovered, two

---

[1]While provision is made for adaptive feedback in data
analysis, usually the possibility of acquiring further data
to illuminate points at issue is not. What we do as
statisticians depends heavily on expectations implied by our
training. While a previous generation of graduates might
have expected to prove theorems, occasionally to test an
isolated hypothesis, and perhaps to teach a new generation
of students to do likewise, the present generation might be
forgiven for believing that their fate is only to explore
"data sets" and speculate on what might or might not explain
them. We must encourage our students to accept the heritage
bestowed by Fisher, who elevated the statistician from an
archivist to an active designer of experiments and hence an
architect and coequal investigator.

circumstances needed to coincide: (a) a potentially
informative experience needed to occur, and (b) the
phenomenon needed to be known about by someone of sufficient
acuity of mind to formulate, and preferably to test, a
possible rule for its future occurrence.

Progress was slow because of the rarity of the two
necessary individual circumstances and the still greater
rarity of their coincidence. Experimental science accel-
erates this learning process by isolating its essence:
potentially informative experiences are deliberately
staged and made to occur in the presence of a trained
investigator. As science has developed, we have learned how
such artificial experiences may be carefully contrived to
isolate questions of interest, how conjectures that are put
forward may be tested, and how residual differences from
what had been expected can be used to modify and improve
initial ideas. So the ordinary process of learning has been
sharpened and accelerated.

The instrument of all learning is the brain - an
incredibly complex structure, the working of which we have
only recently begun to understand. One thing that is clear
is the importance to the brain of models. To appreciate why
this is so, consider how helpless we would be if, each
night, all our memories were eliminated, so that we awoke to
each new day with no past experiences whatever and hence no
models to guide our conduct. In fact, our past experience
is conveniently accumulated in models $M_1, M_2, \ldots, M_i, \ldots$
Some of these models are well established, others less so,
while still others are in the very early stages of creation.
When some new fact or body of facts y_d comes to our
attention, the mind tries to associate this new experience
with an established model. When, as is usual, it succeeds
in doing so, this new knowledge is incorporated in the
appropriate model and can set in train appropriate action.

Obviously, to avoid chaos the brain must be good at
allocating data to an appropriate model and at initiating
the construction of a new model if this should prove to be
necessary. To conduct such business the mind must be able

to deduce what facts could be expected as realizations of a particular model and, more difficult, to induce what model(s) are consonant with particular facts.

Thus, it is concerned with two kinds of inference: (A) Contrasting of new facts y_d with a possible model M: an operation I will characterize by subtraction M − y_d. This process stimulates induction and will be called criticism. (B) Incorporating new facts y_d into an appropriate model: an operation I will characterize by addition M + y_d. This process is entirely deductive and will be called estimation.

I believe then that many of our difficulties arise because, while there is an essential need for two kinds of inference, there seems an inherent propensity among statisticians to seek for only one.

In any case, research which, following the discoveries of Roger Sperry and his associates, has gathered great momentum in the past 25 years shows that the human brain behaves not as a single entity but as two largely separate but cooperating instruments which do two different things (see for example Springer and Deutsch [1981], Blackeslee [1980]).

In most people[2], the left brain is concerned primarily with language and logical deduction, the right brain with images, patterns and inductive processes. The two sides of the brain are joined by millions of connections in the corpus callosum. It is known that the left brain plays a conscious and dominant role while by contrast one may be quite unaware[3] of the working of the right brain.

---

[2] In about one third of left-handed people (about 5% of the population) the roles of the right and left brain are reversed.

[3] For example the apparently instinctive knowledge of what to do and how to do it, enjoyed by the experienced tennis player and by the experienced motorist, comes from the right brain.

The right brain's ability to appreciate[4] patterns in
data y_d and to find patterns in discrepancies $M_i - y_d$
between the data and what might be expected if some
tentative model were true is of great importance in the
search for explanations of data and of discrepant events.
This accomplishment of the right brain of pattern
recognition is of course of enormous consequence in
scientific discovery[5]. However, some check is needed on
its pattern seeking ability, for common experience shows
that some pattern or other can be seen in almost any set of
data or facts[6]. This is the object of diagnostic checks
and tests of fit which, I will argue, require frequentist
theory significance tests for their formal justification.

3. THE THEORY - PRACTICE ITERATION.

It has long been recognized that the learning process
is a motivated iteration between theory and practice. By
practice I mean reality in the form of data or facts. In
this iteration deduction and induction are employed in
alternation. Progress of an investigation is thus evidenced
by a theoretical model, which is not static, but by
appropriate exposure to reality continually evolves until

---

[4]Implicit recognition of the need to stimulate the
remarkable pattern-seeking ability of the right brain is
evidenced by modern emphasis on ingenious plotting devices
in the model formulation/modification phases of
investigation. In particular Chernoff's representation of
multivariate data by faces [1973] and earlier Edgar
Anderson's use of glyphs [1960] direct the right brain to
the recognition problem at which it excels.

[5]Manifestations of the importance to discovery of
unconscious pattern seeking by the right brain have often
been noticed. For example, Beveridge [1950] remarks that
happenings of the following kind are commonplace: a
scientist has mulled over a set of data for many months and
then, at a certain point in time, perhaps on a country walk
when the problem is not being consciously thought about, he
suddenly becomes aware of a solution (model) which explains
these data. This point in time is presumably that at which
the right brain sees fit to let the left brain know what it
has figured out.

[6]See, for example, the King of Heart's rationalization of
the poem brought as evidence in the trial of the Knave of
Hearts in Lewis Carrol's Alice in Wonderland.

some currently satisfactory level of understanding is
reached. At any given stage in a scientific investigation
the current model helps us to appreciate not only what we
know, but what else it may yet be important to find out and
so motivates the collection of new data to illuminate dark
but possibly interesting corners of present knowledge. See
for example Box and Youle [1955], Box [1976], Box, Hunter
and Hunter [1978].

The reader can find illustration of these matters in
his everyday experience, or in the evolution of the plot of
any good mystery novel, as well as in any reasonably honest
account of the events leading to scientific discovery.

Different levels of adaptation

The adaptive iteration we have described produces
change in what we believe about the system being studied,
but it can also produce change in how we study it, and
sometimes even in the objective[7] of the study. This
multiple adaptivity explains the surprising property of
convergence of a process of investigation which at first
appears hopelessly arbitrary. See for example Box [1957].
To appreciate this arbitrariness, suppose that some
scientific problem were being studied by, say, 10
independent sets of investigators, all competent in the
field of endeavor. It is certain that they would start from
different points, conduct the investigation in different
ways, have different initial ideas about which variables
were important, on what scales, and in which transformation. Yet it is perfectly possible that they would all
eventually reach similar conclusions. It is important
to bear this context of multiple iteration in mind

---

[7]If we start out to prospect for silver, we should not
ignore an accidental discovery of gold. For example, one
experimental attempt to find manufacturing conditions giving
greater yield of a particular product failed to find any
such, but did find reaction conditions giving the same yield
with the reaction time halved. This meant that, by
switching to the new manufacturing conditions, throughput
could be doubled, and that a costly, previously planned,
extension of the plant was unnecessary.

when we consider the scientific process and how it relates
to a statistical method.

4. STATISTICAL ESTIMATION AND CRITICISM.

In a recent paper (Box [1980]) a statistical theory was
presented which, it was argued, was consonant with the view
of scientific investigation outlined above. Suppose at the
i^{th} stage of such an investigation a set of assumptions
A_i are tentatively entertained which postulate that to an
adequate approximation, the density function for potential
data y is $p(y|\theta,A_i)$ and the prior distribution for θ
is $p(\theta|A_i)$. Then it was argued that the model M_i should
be defined as the joint distribution of y and θ

$$p(y,\theta|A_i) = p(y|\theta,A_i)p(\theta|A_i) \tag{1}$$

since it is a complete statement of prior tentative belief
at stage i. In these expressions A_i is understood to
indicate all or some of the assumptions in the model
specification at stage i. The model of equation (1) means
to me that current belief about the outcome of contemplated
data acquisition would be calibrated with adequate
approximation by a physical simulation involving appropriate
random sampling from the distributions $p(y|\theta,A_i)$ and
$p(\theta|A_i)$.

The model can also be factored as

$$p(y,\theta|A) = p(\theta|y,A)p(y|A) . \tag{2}$$

The second factor on the right, which can be computed before
any data become available,

$$p(y|A) = \int p(y|\theta,A)p(\theta|A)d\theta \tag{3}$$

is the predictive distribution of the totality of all
possible samples y that could occur if the assumptions
were true.

When an actual data vector y_d becomes available

$$p(y_d,\theta|A) = p(\theta|y_d,A)p(y_d|A) . \tag{4}$$

The first factor on the right is the Bayes' posterior
distribution of θ given y_d

$$p(\theta|y_d,A) \propto p(y_d|\theta,A)p(\theta|A) \tag{5}$$

while the second factor

$$p(\underline{y}_d|A) = \int p(\underline{y}_d|\underline{\theta},A)p(\underline{\theta}|A)d\underline{\theta} \, , \tag{6}$$

is the <u>predictive density</u> associated with the particular data $\underline{y}_d$ actually obtained conditional on the truth of the model and on the data $\underline{y}_d$ having occurred.

The posterior distribution $p(\underline{\theta}|\underline{y}_d,A)$ allows all relevant estimation inferences to be made about $\underline{\theta}$, but this posterior distribution can supply no information about the <u>adequacy</u> of the model. Information on adequacy may be provided, however, by reference of the density $p(\underline{y}_d|A)$ to the predictive reference distribution $p(\underline{y}|A)$ or of the density $p\{g_i(\underline{y}_d)|A\}$ of some relevant checking function $g_i(\underline{y}_d)$ to its predictive distribution and in particular by computing the probabilities

$$\Pr\{p(\underline{y}|A) < p(\underline{y}_d|A)\} \tag{7}$$

and

$$\Pr[p\{g_i(\underline{y})|A\} < p\{g_i(\underline{y}_d)|A\}] \tag{8}$$

Two illustrative examples follow.

4.1. The Binomial Model

As an elementary example, suppose inferences are to be made about the proportion θ of successes in a set of binomial trials.

Suppose n trials are <u>about to be made</u> and assume a beta-distribution prior with mean θ_0. Then

$$p(\theta|A) = [B\{(m\theta_0,m(1-\theta_0)\}]^{-1}\theta^{m\theta_0-1}(1-\theta)^{m(1-\theta_0)-1} \tag{9}$$

$$p(y|\theta,A) = \binom{n}{y}\theta^y(1-\theta)^{n-y} \tag{10}$$

and the predictive distribution is

$$p(y|A)=\binom{n}{y}[B\{m\theta_0,m(1-\theta_0)\}]^{-1}B\{m\theta_0+y,m(1-\theta_0)+n-y\} \tag{11}$$

which may be computed <u>before</u> the data are obtained.

If, now, having performed n trials, there are y_d successes, the likelihood defined up to a multiplicative constant is

$$L(\theta | y_d, A) = \theta^{y_d}(1 - \theta)^{n-y_d} \tag{12}$$

the predictive density is

$$p(y_d|A) = \binom{n}{y_d} [B\{m\theta_0, m(1-\theta_0)\}]^{-1} B\{m\theta_0 + y_d, m(1-\theta_0) + n - y_d\} \tag{13}$$

and the posterior distribution of θ is

$$p(\theta | y_d, A) = [B\{m\theta_0 + y_d, m(1-\theta_0) + n - y_d\}]^{-1} \times$$

$$\theta^{y_d + m\theta_0 - 1}(1 - \theta)^{n - y_d + m(1-\theta_0) - 1}. \tag{14}$$

In the examples of Figures 1 and 2 full lines are used for items available _prior_ to the availability of data y_d and dotted lines for items available only after the data y_d are in hand. Both Figures 1 and 2 illustrate a situation where the prior distribution $p(\theta|A)$ has mean $\theta_0 = 0.2$ and $m = 20$ and we know that $n = 10$ trials are to be performed. Knowing these facts, we can immediately calculate the predictive distribution $p(y|A)$ which is the probability distribution for all possible outcomes from such a model if we suppose the model is true.

When the experiment is actually performed suppose at first, as in Figure 1, that $y_d = 3$ of the trials are successes. The predictive probability $p(3|A)$ associated with this outcome is not unusually small. In fact $\Pr\{p(y|A) < p(3|A)\} = 0.42$ and we have no reason to question the model. Thus for this sample the likelihood $L(\theta|y)$ may reasonably be combined with the prior to produce the posterior distribution shown.

In Figure 2 however it is supposed instead that the outcome is $y_d = 8$ successes so that for this sample $\Pr\{p(y|A) < p(8|A)\} = 0.0013$ and the adequacy of the model, and in particular the adequacy of the prior distribution, is now called into question. Inspection of the figure shows how this agrees with common sense; for in the case illustrated the posterior distribution is unlike either the prior distribution or the likelihood which were combined to obtain it.

G. E. P. Box

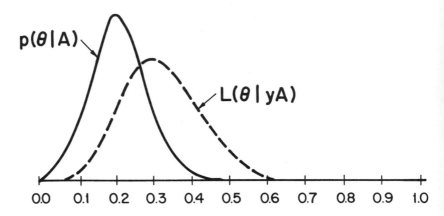

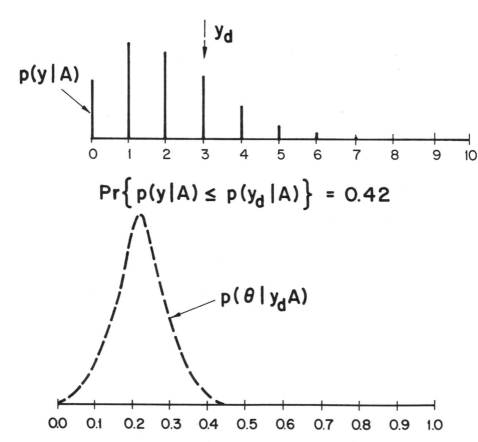

Figure 1.   Prior, likelihood, predictive and posterior
distributions for $n = 10$ Bernoulli trials
with $y_d = 3$ successes.

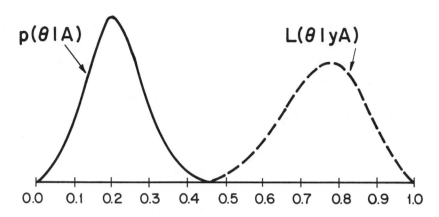

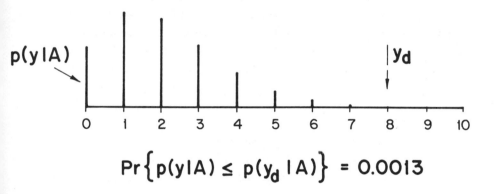

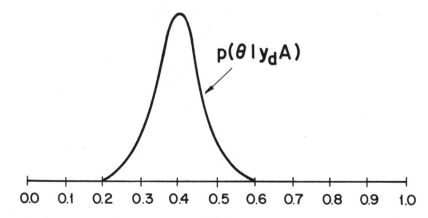

Figure 2. Prior, likelihood, predictive and posterior
distributions for n = 10 Bernoulli trials
with y_d = 8 successes.

Misgivings about the use of Bayes' theorem which some
have expressed in the past are certainly associated with the
possibility of distorting the information coming from the
data by the use of an inappropriate prior distribution.
Without predictive checks, the following objections would
carry great weight:

(a) that nothing in the Bayes' calculation of the
posterior distribution itself could warn of the
incompatibility of the data and the model, and especially
the prior; and

(b) that in complicated examples it would not be so
obvious when this incompatibility occurred.

A case of particular interest occurs when the prior is
sharply centered[8] at its mean value $\theta_0 = 0.2$. This
happens in the above binomial setup when m is made very
large. Then, if the model is unquestioned, the posterior
distribution will be essentially the same as the prior
leading to the conclusion that θ is close to θ_0 whatever
the data. The predictive distribution in this case is
$p(y|\theta_0, A)$, the ordinary binomial sampling distribution,
and the predictive check is the standard binomial
significance test, which can discredit the model with
$\theta = \theta_0 = 0.2$ and hence discredit the application of Bayes'
theorem to this case. This, to my mind, produces the most
satisfactory justification for the standard significance
test.

4.2.   The Normal Linear Model and Ridge Estimators

Another example, discussed in Box [1980], concerns the
normal linear model. In a familiar notation suppose

$$y \sim N(\underset{\sim}{1}\mu + \underset{\sim}{X}\theta, \underset{\sim}{I}_n\sigma^2) \tag{15}$$

with $\underset{\sim}{1}$ a vector of unities and $\underset{\sim}{X}$ of full rank k and
such that $\underset{\sim}{X}'\underset{\sim}{1} = \underset{\sim}{0}$ and suppose that prior densities are
locally approximated by

---

[8] Such a model with a prior sharply centered at $\theta_0 = 0.2$
might be appropriate, for instance, if a trial consisted of
spinning ten times what seemed to be a properly balanced
pentagonal top and counting the number of times the top fell
on a particular segment.

$$\mu \sim N(\mu_0, c^{-1}\sigma^2), \underset{\sim}{\theta} \sim N(\underset{\sim}{\theta}_0, \underset{\sim}{\Gamma}^{-1}\sigma^2), \{\sigma^2/\nu_0 s_0^2\} \sim \chi^{-2}(\nu_0) \tag{16}$$

where $\chi^{-2}(\nu_0)$ refers to the inverted χ^2 distribtuion with ν_0 degrees of freedom and μ and $\underset{\sim}{\theta}$ independent conditional on σ^2.

Given a sample $\underset{\sim}{y}_d$, special interest attaches to $\underset{\sim}{\theta}$ and σ^2 which, given the assumptions, are estimated by $p(\underset{\sim}{\theta}, \sigma^2 | \underset{\sim}{y}_d, A)$ with marginal distributions

$$p(\underset{\sim}{\theta} | \underset{\sim}{y}_d, A) \propto \{1 + \frac{(\underset{\sim}{\theta} - \bar{\underset{\sim}{\theta}}_d)'(\underset{\sim}{X}'\underset{\sim}{X} + \underset{\sim}{\Gamma})(\underset{\sim}{\theta} - \bar{\underset{\sim}{\theta}}_d)}{(n + \nu_0)\hat{\sigma}_d^2}\}^{-\frac{1}{2}(n+\nu_0+k)} \tag{17}$$

$$p(\sigma^2 | \underset{\sim}{y}_d, A) \propto \sigma^{-(n+\nu_0+2)} \exp\{-\frac{1}{2}(n+\nu_0)\hat{\sigma}_d^2/\sigma^2\} \tag{18}$$

with

$$\bar{\underset{\sim}{\theta}}_d = (\underset{\sim}{X}'\underset{\sim}{X} + \underset{\sim}{\Gamma})^{-1}(\underset{\sim}{X}'\underset{\sim}{X}\hat{\underset{\sim}{\theta}}_d + \underset{\sim}{\Gamma}\underset{\sim}{\theta}_0) ,$$

$$\hat{\underset{\sim}{\theta}}_d = (\underset{\sim}{X}'\underset{\sim}{X})^{-1}\underset{\sim}{X}'\underset{\sim}{y}_d, \qquad \nu = n - k - 1 , \tag{19}$$

$$(n+\nu_0)\hat{\sigma}_d^2 = \nu s_d^2 + \nu_0 s_0^2 + (\hat{\underset{\sim}{\theta}}_d - \underset{\sim}{\theta}_0)'\{(\underset{\sim}{X}'\underset{\sim}{X})^{-1} + \underset{\sim}{\Gamma}^{-1}\}^{-1}(\hat{\underset{\sim}{\theta}}_d - \underset{\sim}{\theta}_0)$$

$$+ (n^{-1} + c^{-1})^{-1}(\bar{y} - \mu_0)^2 .$$

and

$$s^2 = \{\underset{\sim}{I} - \underset{\sim}{X}(\underset{\sim}{X}'\underset{\sim}{X})^{-1}\underset{\sim}{X}'\}\underset{\sim}{y}, \quad s_d^2 = \{\underset{\sim}{I} - \underset{\sim}{X}(\underset{\sim}{X}'\underset{\sim}{X})^{-1}\underset{\sim}{X}'\}\underset{\sim}{y}_d . \tag{20}$$

Now let

$$s_p^2 = (\nu + \nu_0)^{-1}(\nu s^2 + \nu_0 s_0^2) \quad \text{and}$$

$$s_{pd}^2 = (\nu + \nu_0)^{-1}(\nu s_d^2 + \nu_0 s_0^2) . \tag{21}$$

Then the joint predictive distribution can be factored into independent components for $(\hat{\underset{\sim}{\theta}} - \underset{\sim}{\theta}_0)/s_p$, s^2, and $\nu - 1$ angular elements of the standarized residuals. A predictive check based on the first of these factors

$$\Pr\{p((\hat{\underset{\sim}{\theta}} - \underset{\sim}{\theta}_0)/s_p | A) < p((\hat{\underset{\sim}{\theta}}_d - \underset{\sim}{\theta}_0)/s_{pd} | A)\}$$

$$= \Pr\{F_{k, \nu+\nu_0} > \frac{(\hat{\underset{\sim}{\theta}}_d - \underset{\sim}{\theta}_0)'\{(\underset{\sim}{X}'\underset{\sim}{X})^{-1} + \underset{\sim}{\Gamma}^{-1}\}^{-1}(\hat{\underset{\sim}{\theta}}_d - \underset{\sim}{\theta}_0)}{k s_{pd}^2}\} \tag{22}$$

is the standard analysis of variance check for compatibility
of two estimates $\hat{\theta}_d$ and $\hat{\theta}_0$ and was earlier proposed as a
check for compatibility of prior and sample information by
Theil [1963].

Now suppose the X matrix to be in correlation form
and assume $\theta_0 = 0, \Gamma = I_k \gamma_0, \nu_0 \to 0$ so that $s_p^2 \to s^2$. Then
the estimates $\bar{\theta}_d$ are the ridge estimators of Hoerl and
Kennard [1970] which, given the assumptions, appropriately
combine information from the prior with information from the
data. The predictive check (22) now yields

$$\alpha = \Pr\{F_{k,\nu} > \frac{\hat{\theta}_d'\{(X'X)^{-1} + I\gamma_0^{-1}\}^{-1}\hat{\theta}_d}{ks_d^2}\} \tag{23}$$

allowing any choice of γ_0 to be criticized.

For example, in their original analysis of the data of
Gorman and Toman [1966], Hoerl and Kennard [1970] chose a
value $\gamma_0 = 0.25$. But substitution of this value in (23)
yields $\alpha = \Pr\{F_{10,25} > 3.59\} < 0.01$ which discredits this
choice.

One can see for these examples how the two functions of
criticism and estimation are performed by the predictive
check on the one hand and the Bayesian posterior
distribution on the other.

Thus consider the ridge (Bayes' mean) estimator of the
second example. This estimator is a linear combination of
the least squares estimate $\hat{\theta}$ and the prior mean θ_0, with
weights supplied by the appropriate information matrices,
and with covariance matrix obtained by inverting the <u>sum</u> of
these <u>information matrices</u>. Assuming the data to be a
realization of the model, this is the appropriate way of
combining the two sources of information.

The predictive check, on the other hand, <u>contrasts</u> the
values $\hat{\theta}$ and θ_0 with a dispersion matrix obtained by
appropriately <u>summing</u> the two <u>dispersion matrices</u>.

The <u>combination</u> of information from the prior and
likelihood into the posterior distribution and the
<u>contrasting</u> of these two sources of information in the
predictive distribution is equally clear in the binomial

example and especially in its appropriate normal approximation.

5. SOME OBJECTIONS CONSIDERED

A recapitulation of the argument and a consideration of some objections is considered in this section.

5.1. Essential elements of the argument

A. Scientific investigation is an iterative process in which the model is not static but is continually evolving. At a given stage the nature of the uncertainties in a model directs the acquisition of further data, whether by choosing the design of an experiment or sample survey, or by motivating a search of a library or data bank. At, say, the ith stage of an investigation all current structural assumptions A_i, including those about the prior, must be thought of, not as being true, but rather as being subjective guesses which at this particular stage of the investigation are worth entertaining. It is consistent with this attitude that when data y_d become available checks need to be applied to assess consonance with A_i.

B. The statistical <u>model</u> at the ith stage of the investigation should be defined as the joint distribution of y and θ given the assumptions A_i

$$p(y,\theta|A_i) = p(y|\theta,A_i)p(\theta|A_i) \ . \tag{24}$$

C. Not one but two distinct kinds of inference are involved within the iterative process: <u>criticism</u> in which the appropriateness of regarding data y_d as a realization of a particular model M is questioned; <u>estimation</u> in which the consequences of the assumption that data y_d are a realization of a model M are made manifest.

This criticism-estimation dichotomy is characterized mathematically by the factorization of the model realization $p(y_d,\theta|A_i)$ into the predictive density $p(y_d|A_i)$ and the posterior distribution $p(\theta|y_d,A_i)$. The predictive distribution $p(y|A_i)$ provides a reference distribution for $p(y_d|A_i)$. Similarly the predictive distribution $p\{g(y)|A_i\}$ of any checking function $g(y)$ provides a reference distribution of the corresponding predictive

density $p\{g(\underline{y}_d|A_i\}$. Unusually small values of this density
suggest that the current model is open to question.

D. If we are satisfied with the adequacy of the
assumptions A_i then the posterior distribution
$p(\underline{\theta}|\underline{y}_d,A_i)$ allows for complete estimation of $\underline{\theta}$ and no
other procedures of <u>estimation</u> are relevant. In particular,
therefore, insofar as shrinkage, ridge and robust estimators
are useful, they ought to be direct consequences of an
appropriate model and should not need the invocation of
extraneous considerations such as minimization of mean
square error.

Objections. Numbered to correspond with the various
elements of the argument are responses to some objections
that have been, or might be, raised.

A(i) <u>Iterative investigation</u>? Some would protest that
their own statistical experience is not with iterative
investigation but with a single set of data to be analyzed,
or a single design to be laid out and the results
elucidated.

Many circumstances where the statistician has been
involved in a "one-shot" analysis rather than an iterative
partnership, ought not to have happened. Such involvement
frequently occurs when the statistician has been drafted as
a last resort, all other attempts to make sense of the data
having failed. At this point data gathering will usually
have been completed and there is no chance of influencing
the course of the study. Statisticians whose training has
not exposed them to the overriding importance of
experimental design are most likely to acquiesce in this
situation, or even to think of it as normal, and thus to
encourage its continuance.

The statistician who has cooperated in the design of a
<u>single</u> experiment which he analyzes is somewhat better
off. However one-shot designs are often inappropriate
also. Underlying most investigations is a budget, stated or
unstated, of time and/or money that can reasonably be
expended. Sometimes this latent budget is not adequate to
the goal of the investigation, but, for purposes of
discussion, let us suppose that it is. Then if a

sequential/iterative approach is possible it would usually
be quite inappropriate to plan the whole investigation at
the beginning in one large design. This is because the
results from a first design will almost invariably supply
new and often unexpected information about choice of
variables, metrics, transformations, regions of operability,
unexpected side-effects, and so forth, which will vitally
influence the course of the investigation and the nature of
the next experimental arrangement. A rough working rule is
that not more than 25% of the time-and-money budget should
be spent on the first design. Because large designs can in
a limited theoretical sense be more efficient it is a common
mistake not to take advantage of the iterative option when
it is available. Instances have occurred of experimenters
regretting that they were persuaded by an inexperienced
statistician to perform a large "all inclusive" design where
an adaptive strategy would have been much better.   In
particular, it is likely that many of the runs from such
"all-embracing" designs, will turn out to be noninformative
because their structure was decided when least was known
about the problem.

Scientific iteration is strikingly exemplified in
response surface studies (see, for example, Box and Wilson
[1951], Box [1954], Box and Youle [1955]). In particular
methods such as steepest ascent and canonical analysis can
lead to exploration of new regions of the experimental
space, requiring elucidation by new designs which, in turn,
can lead to the use of models of higher levels of
sophistication. Although in these examples the necessity
for such an iterative theory is most obvious, it clearly
exists much more generally, for example in investigations
employing sequences of orthodox experimental designs and to
many applications of regression analysis. It has sometimes
been suggested that agricultural field trials are not
sequential but of course this is not so; only the time frame
is longer. Obviously what is learned from one year's work
is used to design the next year's experiments.

However I agree that there are some more convincing
exceptions. For example, a definitive trial which is

intended to settle a controversy such as a test of the
effectiveness of Laetrile as a cure for cancer. Also the
iteration can be very slow. For example, in trials on the
weathering of paints, each phase can take from 5-10 years.

A(ii) Subjective probability? The view of the process
of scientific investigation as one of model evolution has
consequences concerning subjective probabilities. An
objection to a subjectivist position is that in presenting
the final results of our investigation, we need to convince
the outside world that we have really reached the conclusion
that we say we have. It is argued that, for this purpose,
subjective probabilities are useless. However I believe
that the confirmatory stage of an iterative investigation,
when it is to be demonstrated that the final destination
reached is where it is claimed to be, will typically occupy,
perhaps, only the last 5 per cent of the experimental
effort. The other 95 per cent - the wandering journey that
has finally led to that destination - involves, as I have
said, many heroic subjective choices (what variables? what
levels? which scales? etc., etc.) at every stage. Since
there is no way to avoid these subjective choices which are
a major determinant of success why should we fuss over
subjective probability?

Of course, the last 5 per cent of the investigation
occurs when most of the problems have been cleared up and we
know most about the model. It is this rather minor part of
the process of investigation that has been emphasized by
hypothesis testers and decision theorists. The resultant
magnification of the importance of formal hypothesis tests
has inadvertently led to underestimation by scientists of
the area in which statistical methods can be of value and to
a wide misunderstanding of their purpose. This is often
evidenced in particular by the attitudes to statistics of
editors and referees of journals in the social, medical and
biological sciences.

B(i) The Statistical Model? The statistical model has
sometimes been thought of as the density function $p(y|\theta,A)$
rather than the joint density $p(y,\theta|A)$ which reflects the

influence of the prior. However only the latter form
contains all currently entertained beliefs about y and
θ. It seems quite impossible to separate prior belief from
assumptions about model structure. This is evidenced by the
fact that assumptions are frequently interchangeable between
the density $p(y|\theta)$ and the prior $p(\theta)$. As an elementary
example, suppose that among the parameters $\theta = (\phi, \beta)$ of a
class of distributions β is a shape parameter such that
$p(y|\phi, \beta_0)$ is the normal density. Then it may be
convenient, for example in studies of robustness, to define
a normal distribution by writing the more general density
$p(y|\phi, \beta)$ with an associated prior for β which can be
concentrated at $\beta = \beta_0$. The element specifying normality
which in the usual formulation is contained in the density
$p(y, \theta)$ is thus transferred to the prior $p(\theta)$.

B(ii) Do we need a prior? Another objection to the
proposed formulation of the model is the standard protest of
non-Bayesians concerning the introduction of any prior
distribution as an unnecessary and arbitrary element.
However, recent history has shown that it is the omission in
sampling theory, rather than the inclusion in Bayesian
analysis, of an appropriate prior distribution, that leads
to trouble.

For instance Stein's result [1955] concerning the
inadmissibility of the vector of sample averages as an
estimate of the mean of a multivariate normal distribution
is well known. But consider its practical implication for,
say, an experiment resulting in a one-way analysis of
variance. Such an experiment could make sense when it is
conducted to compare, for example, the levels of infestation
of k different varieties of wheat, or the numbers of eggs
laid by k different breeds of chickens or the yields of
k successive batches of chemical; in general, that is,
when a priori we expect similarities of one kind or another
between the entities compared. But clearly, if similarities
are in mind, they ought not to be denied by the form of the
model. They are so denied by the improper prior which
produces as Bayesian means the sample averages, which are in
turn the orthodox estimates from sampling theory.

Now the reason that k wheat varieties, k chicken breeds or k batch yields are being jointly considered is because they are, in one sense or another, comparable. The presence of a specific form of prior distribution allows the investigator to incorporate in the model precisely the kind of similarities he wishes to entertain. Thus in the comparison of varieties of wheat or of breeds of chicken it might well be appropriate to consider the variety means as randomly sampled from some prior super-population and, as is well known, this can produce the standard shrinkage estimators as Bayesian means (Lindley [1965], Box and Tiao [1968], Lindley and Smith [1972]). But notice that such a model is likely to be quite inappropriate for the yields of k successive batches of chemical. These mean yields might much more reasonably be regarded as a sequence from some autocorrelated time series. A prior which reflected this concept led Tiao and Ali [1971] to functions for the Bayesian means which are quite different from the orthodox shrinkage estimators.

In summary, then, both sampling theory and Bayes theory can rationalize the use of shrinkage estimators, and the fact that the former does so merely on the basis of reduction of mean square error with no overt use of a prior distribution, at first seems an advantage. However, only the explicit inclusion of a prior distribution, which sensibly describes the situation we wish to entertain, can tell us what is the appropriate function to consider, and avoid the manifest absurdities which seem inherent in the sampling theory approach which implies, for example, that we can improve estimates by considering as one group varieties of wheat, breeds of chicken, and batches of chemical.

C(i)   Is there an iterative interplay between criticism
        and estimation? A good example of the iterative
interplay between criticism and estimation is seen in
parametric time series model building as described for
example by Box and Jenkins [1970]. Critical inspection of
the plotted time series and of the corresponding plotted
autocorrelation function, and other functions derivable from
it, together with their rough limits of error, can suggest a

model specification and in particular a parametric model.
Temporarily behaving as if we believed this specification,
we may now _estimate_ the parameters of the time series model
by their Bayesian posterior distribution (which, for samples
of the size usually employed, is sufficiently well indicated
by the likelihood). The residuals from the fitted model are
now similarly _critically examined_, which can lead to
respecification of the model, and so on. Systematic
liquidation of serial dependence brought about by such an
iteration can eventually produce a parametric time series
model; that is a linear filter which approximately
transforms the time series to a white noise series. Anyone
who carries through this process must be aware of the very
different nature of the two inferential processes of
criticism and estimation which are used in alternation in
each iterative cycle.

C(ii)   Why can't all criticism be done using Bayes
        posterior analysis?

It is sometimes argued that model checking can always
be performed as follows: let $A_1, A_2, \ldots, A_k$ be alternative
assumptions; then the computation of

$$p(A_i|y) = \frac{p(y|A_i)p(A_i)}{\sum\limits_{j=1}^{k} p(y|A_j)p(A_j)} \qquad (i = 1, 2, \ldots, k) \qquad (25)$$

yields the probabilities for the various sets of
assumptions.

The difficulty with this approach is that by supposing
all possible sets of assumptions known _a priori_ it
discredits the possibility of new discovery. But new
discovery is, after all, the most important object of the
scientific process.

At first, it might be thought that the use of (25) is
not misleading, since it correctly assesses the _relative_
plausibility of the models considered. But in practice this
would seem of little comfort. For example suppose that
only $k = 3$ models are currently regarded as possible, and
that having collected some data the posterior probabilities
$p(A_i|y)$ are 0.001, 0.001, 0.998 $(i = 1,2,3)$. Although

in relation to these particular alternatives $p(A_3|y)$ is
overwhelmingly large this does not necessarily imply that in
the real world assumptions A_3 could be safely adopted.
For, suppose unknown to the investigator, a fourth
possibility A_4 exists which given the data is a thousand
times more probable than the group of assumptions previously
considered. Then, if that model had been included, the
probabilities would be 0.000,001, 0.000,001, 0.000,998, and
0.999,000.

Furthermore, <u>in ignorance of</u> A_4 it is highly likely
that a study of the components of the predictive
distribution $p(y|A_3)$ and in particular of the residuals,
could (a) have shown that A_3 was not acceptable and (b)
have provided clues as <u>to the identity of</u> A_4. The
objective of good science must be to conjure into existence
what has <u>not</u> been contemplated previously. A Bayesian
theory which excludes this possibility subverts the
principle aim of scientific investigation.

More generally, the possibility that there are more
than one set of assumptions that may be considered, merely
extends the definition of the <u>model</u> to

$$p(y,\theta,A_j) = p(y|\theta A_j)p(\theta|A_j)p(A_j) \qquad (j = 1,2,\ldots,k)$$

which in turn will yield a predictive distribution. In a
situation when this more general model is inadequate a
mechanical use of Bayes theorem could produce a misleading
analysis, while suitable inspection of predictive checks
could have demonstrated, on a sampling theory argument, that
the global model was almost certainly wrong and could have
indicated possible remedies.[9]

C(iii) <u>An abrogation of the likelihood principle</u>? The
likelihood principle holds, of course, for the estimation
aspect of inference in which the model is temporarily
assumed true. However it is inapplicable to the criticism
process in which the model is regarded as in doubt.

---

[9]I am grateful to Dr. Michael Titterington for pointing out that in
discriminant analysis the atypicality indices of Aitchison and Aitken
[1976] use similar ideas.

If the assumptions A are supposed true, the likelihood function contains all the information about $\underset{\sim}{\theta}$ coming from the particular observed data vector y_d. When combined with the prior distribution for $\underset{\sim}{\theta}$ it therefore tells all we can know about $\underset{\sim}{\theta}$ given y_d and A. In such a case the predictive density $p(y_d|A)$ can tell us nothing we have not already assumed to be true, and will fall within a given interval with precisely the frequency forecast by the predictive distribution. When the assumptions are regarded as possibly false, however, this will no longer be true and information about model inadequacy can be supplied by considering the density $p(y_d|A)$ in relation to $p(y|A)$. Thus for the Normal linear model, the distribution of residuals contains no information if the model is true, but provides the reference against which standard residual checks, graphical and otherwise, are made on the supposition that it may be importantly false.

In the criticism phase we are considering whether, given A, the sample y_d is likely to have occurred at all. To do this we <u>must</u> consider it in relation to the <u>other</u> samples that could have occurred but did not.

For instance in the Bernoulli trial example, had we sampled until we had r successes rather than until we had n trials, then the likelihood, and, for a fixed prior, the posterior distribution, would have been unaffected, but the predictive check would (appropriately) have been somewhat different because the appropriate reference set supplied by $p(y|A)$ would be different.

C(iv)   <u>How do you choose the significance level?</u>

It has been argued that if significance tests are to be employed to check the model, then it is necessary to state in advance the level of significance α which is to be used and that no rational basis exists for making such a choice.

While I believe the ultimate justification of model checking is the reference of the checking function to its appropriate predictive distribution, the examples I have given to illustrate the predictive check may have given a misleading idea of the formality with which this should be done. In practice the predictive check is not intended as a

formal test in the Neyman-Pearson sense but rather as a
rough assessment of signal to noise ratio. It is needed to
see which indications might be worth pursuing. In practice
model checks are frequently graphical, appealing as they
should to the pattern recognition capability of the right
brain. Examples are to be found in the Normal probability
plots for factorial effects and residuals advocated by
Daniel [1959], Atkinson [1973] and Cook [1977]. Because
spurious patterns may often be seen in noisy data some rough
reference of the pattern to its noise level is needed.

D. As might be expected the mistaken search for a
single principle of inference has resulted in two kinds of
incongruity:

> attempts to base estimation on sampling theory, using
> point estimates and confidence intervals; and
> attempts to base criticism and hypothesis testing
> entirely on Bayesian theory.

The present proposals exclude both these possibilities .

Concerning estimation, we will not here recapitulate
the usual objections to confidence intervals and point
estimates but will consider the latter in relation to
shrinkage estimators, ridge estimators, and robust
estimators. From the traditional sampling theory point of
view these estimators have been justified on the ground that
they have smaller mean square error then traditional
estimators. But from a Bayesian viewpoint, they come about
as a direct result of employing a credible rather than an
incredible model. The Bayes' approach provides some
assurance against incredibility since it requires that all
assumptions of the model be clearly visible and available
for criticism.

For illustration, emphasized below by underlining, are
the assumptions that would be needed for a Bayesian
justification of standard linear least squares. We must
postulate not only the model

$$y_u = \underset{\sim}{\theta}' \underset{\sim}{x}_u + e_u \qquad u = 1, 2, \ldots, n \tag{26}$$

with the e_u's <u>independently and normally</u> distributed with

<u>constant variance</u> σ^2 , but also postulate an <u>improper prior</u>
for $\underset{\sim}{\theta}$ and σ^2 .

(a) Consider first the choice of prior. As was
pointed out by Anscombe [1963], if we use a measure such as
 $\underset{\sim}{\theta}'\underset{\sim}{\theta}$ to gauge the size of the parameters, a locally flat
prior for $\underset{\sim}{\theta}$ implies that the larger is the size measure
 $\underset{\sim}{\theta}'\underset{\sim}{\theta}$ the more probable it becomes. The model is thus
incredible. From a Bayesian viewpoint shrinkage and ridge
estimators imply more credible choices of the model, which,
even though approximate are not incredible.

(b) For data collected serially (in particular, for
much economic data) the assumption of error independence in
equation (26) is equally incredible and again its violation
can lead to erroneous conclusions. See for example Coen,
Gomme and Kendall [1969] and Box and Newbold [1971].

(c) The assumption that the specification in (26) is
necessarily appropriate for <u>every</u> subscript $u = 1,2,...,n$
is surely incredible. For it implies that the experimenter's answer to the question "Could there be a small
probability (such as 0.001) that any one of the experimental
runs was unwittingly misconducted?" is "No; that probability
is exactly zero."

So far as the last assumption is concerned a more
credible model considered by Jeffreys [1932], Dixon
[1953], Tukey [1960] and Box and Tiao [1968] supposes that
the error e is distributed as a mixture of Normal
distributions

$$p(e|\underset{\sim}{\theta},\sigma) = (1 - \alpha)f(e|\underset{\sim}{\theta},\sigma) + \alpha f(e|\underset{\sim}{\theta},k\sigma) . \qquad (27)$$

This model was used by Bailey and Box [1980] to estimate the
15 coefficients in the fitted model

$$y = \beta_0 + \sum_{i=1}^{4} \beta_i x_i + \sum_{i=1}^{4} \sum_{j>i}^{4} \beta_{ij} x_i x_j + \sum_{i=1}^{4} \beta_{ii} x_i^2 + e \qquad (28)$$

using data from a balanced incomplete 3^4 factorial design.
Table 1 shows some of their Bayes' estimates (marginal means
and standard deviations of the posterior distribution). For
simplicity, only a few of the coefficients are shown; the
behaviour of the others is similar. Table 1a uses data from

(a) BOX BEHNKEN DATA ONE OR TWO SUSPECT VALUES

	Least Squares	Robust				
ε	Zero	.001	.005	.010	.015	.020
α	Zero	.005	.024	.048	.070	.091
Estimates						
β_4	-3.7	-3.2	-3.1	-3.1	-3.1	-3.1
	(.5)	(.3)	(.2)	(.2)	(.2)	(.2)
β_{44}	-2.6	-3.0	-3.1	-3.1	-3.1	-3.1
	(.7)	(.4)	(.4)	(.4)	(.4)	(.4)
β_{13}	-3.8	-3.8	-3.8	-3.8	-3.8	-3.8
	(.8)	(.4)	(.4)	(.4)	(.3)	(.3)
β_{14}	1.0	-.5	-.5	-.5	-.5	-.5
	(.8)	(.9)	(.9)	(.9)	(.9)	(.9)

(b) BACON DATA NO SUSPECT VALUES

	Least Squares	Robust				
ε	Zero	.001	.005	.010	.015	.020
α	Zero	.005	.024	.048	.070	.091
Estimates						
β_4	4.7	4.7	4.7	4.7	4.7	4.7
	(.3)	(.3)	(.3)	(.3)	(.3)	(.3)
β_{44}	.9	.9	.9	.9	.9	.9
	(.5)	(.5)	(.5)	(.5)	(.5)	(.5)
β_{13}	.8	.8	.8	.8	.8	.8
	(.6)	(.6)	(.5)	(.5)	(.5)	(.5)
β_{14}	-.4	-.4	-.4	-.4	-.4	-.4
	(.6)	(.6)	(.6)	(.6)	(.7)	(.7)

Table 1. Bayesian means with standard deviations (in parentheses) for selected coefficients using various values of (ε, α) in the contaminated model (with $k = 5$).

a paper by Box and Behnken [1960]. These data (see Figure 3)
apparently contain a single bad value (y_{10}), with a small
possibility of a second bad value (y_{13}). Table 1b shows
the same analysis for a second set of data arising from the
same design and published by Bacon [1970], which (see Figure
4) appears to contain no bad values. It was shown by Chen
and Box [1979] that for $k > 5$ the posterior distribution
of β is mainly a function of the single parameter
$\varepsilon = \alpha/(1-\alpha)k$ and the results obtained for $k = 5$ are
labelled in terms of ε as well as α. The analysis is
based on locally noninformative priors on β and on $\log \sigma$
so that the estimates in the first columns of the tables
$(\varepsilon = \alpha = 0)$ are ordinary least squares estimates. The
important point to notice is that for the first set of data
which appears to contain one or two bad values, a major
change away from the least squares estimates can occur as
soon as there is even a slight hint $(\varepsilon = 0.001, \alpha = 0.005)$
of the possibility of contamination. The estimates then
remain remarkably stable for widely different values of ε
over a plausible range.[10] But for the second set (Bacon's
data), which appears to contain no bad values, scarcely any
change occurs at all as ε is changed.

It has been objected that while the Normal model is
inadequate, the contaminated model (27) may be equally so,
and that "therefore" we are better off using ad hoc robust
procedures such as have been recommended by Tukey and others
and justified on the basis of their sampling properties.
This argument loses force, however, since it can be shown by
elementary examples (Chen and Box [1979], Box [1980]) that
the effect of the Bayes' analysis is also to produce
downweighting of the observations with downweighting
functions very similar to those proposed by the
empiricists. However, the Bayes' analysis has the advantage
of being based on a visible model which is itself open to
criticism and has greater adaptivity, doing nothing to

---

[10] They are however (see reply to the discussion of Box [1980])
considerably different from estimates obtained by omitting the suspect
observation and using ordinary least squares.

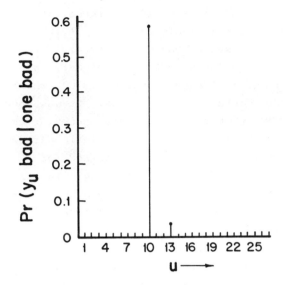

Figure 3.   Posterior probability that y_u is bad given that one observation is bad (Box-Behnken data).

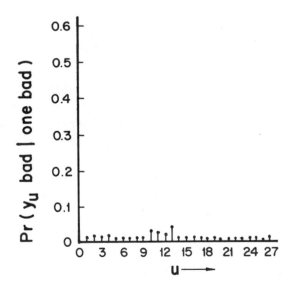

Figure 4.   Posterior probability that y_u is bad given that one observation is bad (Bacon data).

samples that look normal, and reserving robustification for samples that do not. A further advantage of the present point of view is that when an outlier occurs, while the posterior distribution will discount it, the predictive distribution will emphasize it, so that the fact that a discrepancy has occurred is not lost sight of.

CONCLUSION.

In summary I believe that scientific method employs and requires not one, but two kinds of inference - criticism and estimation; once this is understood the statistical advances made in recent years in Bayesian methods, data analysis, robust and shrinkage estimators can be seen as a cohesive whole.

REFERENCES

1. Aitchison, J. and C. G. G. Aitken (1976), "Multivariate Binary Discrimination by the Kernel Method", Biometrika, 63, 413-420.

2. Anderson, E. (1960), "A Semigraphical Method for the Analysis of Complex Problems", Technometrics, 2, 387-391.

3. Anscombe, F. J. (1963), "Bayesian Inference Concerning Many Parameters with Reference to Supersaturated Designs", Bull. Int. Stat. Inst. 40, 721-733.

4. Atkinson, A. (1973), "Testing Transformations to Normality", J. Royal Statis. Soc. B, 35, 473-479.

5. Bacon, D. W. (1970), "Making the Most of a One-shot Experiment", Industrial and Engineering Chem., 62 (7), 27-34.

6. Beveridge (1950), The Art of Scientific Investigation, New York: Vintage Books.

7. Blackeslee, T. R. (1980), The Right Brain, Garden City, New York: Anchor Press/Doubleday.

8. Bailey, S. P. and G. E. P. Box (1980), "The Duality of Diagnostic Checking and Robustification in Model Building: Some Considerations and Examples", Technical Summary Report #2086, Mathematics Research Center, University of Wisconsin-Madison.

9.   Box, G. E. P. (1954), "The Exploration and Exploitation
     of Response Surfaces: Some General Considerations and
     Examples", Biometrics, 10, 16-60.

10. Box, G. E. P. (1957), "Integration of Techniques in
     Process Development", Transactions of the 11th Annual
     Convention of the American Society for Quality Control.

11. Box, G. E. P. (1976), "Science and Statistics",
     J. Amer. Statis. Assoc., 71, 791-799.

12. Box, G. E. P. (1980), "Sampling and Bayes' Inference in
     Scientific Modelling and Robustness", J. Royal Statis.
     Soc. A, 143, 383-430 (with discussion).

13. Box, G. E. P. (1982), "Choice of Response Surface
     Design and Alphabetic Optimality", Yates Volume.

14. Box, G. E. P. and D. W. Behnken (1960), "Some New
     Three-level Designs for the Study of Quantitative
     Variables", Technometrics, 2, 455-475.

15. Box, G. E. P., W. G. Hunter and J. S. Hunter (1978),
     Statistics For Experimenters, New York: John Wiley and
     Sons.

16. Box, G. E. P. and G. M. Jenkins (1970), Time Series
     Analysis: Forecasting and Control, San Francisco:
     Holden-Day.

17. Box, G. E. P. and P. Newbold (1971), "Some Comments on
     a Paper of Coen, Gomme and Kendall", J. Royal Statis.
     Soc. A, 134, 229-240.

18. Box, G. E. P. and G. C. Tiao (1968), "A Bayesian
     Approach to Some Outlier Problems", Biometrika, 55,
     119-129.

19. Box, G. E. P. and G. C. Tiao (1968), "Bayesian
     Estimation of Means for the Random Effect Model", J.
     Amer. Statis. Assoc., 63, 174-181.

20. Box, G. E. P. and K. B. Wilson (1951), "On the
     Experimental Attainment of Optimal Conditions", J.
     Royal Statis. Soc. B, 13, 1-45 (with discussion).

21. Box, G. E. P. and P. V. Youle (1955), "The Exploration
     and Exploitation of Response Surfaces: An Example of
     the Link Between the Fitted Surface and the Basic
     Mechanism of the System", Biometrics, 11, 287-323.

22. Chen, G. G. and G. E. P. Box (1979), "Further Study of Robustification via a Bayesian Approach", Technical Summary Report No. 1998, Mathematics Research Center, University of Wisconsin-Madison.

23. Chernoff, H. (1973), "The Use of Faces to Represent Points in k-Dimensional Space Graphically", J. Amer. Statis. Assoc., 68, 361-368.

24. Coen, P. G., E. D. Gomme and M. G. Kendall (1969), "Lagged Relationships in Economic Forecasting", J. Royal Statis. Soc. A, 132, 133-152.

25. Cook, R. D. (1977), "Detection of Influential Observations in Linear Regression", Technometrics, 19, 15-18.

26. Daniel, C. (1959), "Use of Half-normal Plots in Interpreting Factorial Two-level Experiments", Technometrics, 1, 311-341.

27. Dixon, W. J. (1953), "Processing Data for Outliers", Biometrics, 9, 74-89.

28. Fisher, R. A. (1956), Statistical Methods and Scientific Inference, Edinburgh: Oliver and Boyd.

29. Gorman, J. W. and R. J. Toman (1966), "Selection of Variables for Fitting Equations to Data", Technometrics, 8, 27-51.

30. Hoerl, A. E. and R. W. Kennard (1970), "Ridge Regression: Applications to Non-orthogonal Problems", Technometrics 12, 69-82.

31. Jeffreys, H. (1932), "An Alternative to the Rejection of Observations", Proc. Royal Society, A, CXXXVII, 78-87.

32. Kiefer, J. (1959), Discussion in "Optimum Experimental Designs", J. Royal Statis. Soc. B, 21, 272-319.

33. Lindley, D. V. (1965), Introduction to Probability and Statistics from a Bayesian Viewpoint, Part 2, Inference, Cambridge University Press.

34. Lindley, D. V. and A. F. M. Smith (1972), "Bayes Estimates for the Linear Model", J. Royal Statis. Soc. B, 34, 1-41 (with discussion).

35. Springer, S. P. and G. Deutsch, Left Brain, Right Brain, San Francisco: W. H. Freeman.

36. Stein, C. (1955), "Inadmissibility of the Usual
    Estimator for the Mean of a Multivariate Normal
    Distribution", Proceedings of the Third Berkeley
    Symposium, 197-206.

37. Theil, H. (1963), "On the Use of Incomplete Prior
    Information in Regression Analysis", J. Amer. Statis.
    Assoc., 58, 401-414.

38. Tiao, G. C. and M. M. Ali (1971), "Analysis of
    Correlated Random Effects Linear Model with Two Random
    Components", Biometrika, 58, 37-52.

39. Tukey, J. W. (1960), "A Survey of Sampling from
    Contaminated Distributions", in Contributions to
    Probability and Statistics: Essays in Honor of Harold
    Hotelling, 448-485, Stanford: Stanford University
    Press.

Department of Statistics and
Mathematics Research Center
University of Wisconsin-Madison
Madison, WI 53706

Can Frequentist Inferences Be Very Wrong?
A Conditional "Yes"

David V. Hinkley

"It is a capital mistake to theorize before one has data"
Sherlock Holmes, in Scandal in Bohemia

1. INTRODUCTION

The major operational difference between Bayesian and
frequentist inferences is that in the latter one must choose
a reference set for the sample, in order to obtain inferential probabilities. It is our thesis that in the matter of
choosing a reference set, Sherlock Holmes was right, and
that many frequentist inferences are inadequate because of
erroneous choices made prior to the experiment.

It could not be argued with any conviction that the
mathematization of statistics was other than very beneficial.
The introduction of mathematics permitted logical development of the ideas of sufficiency, efficiency, hypothesis
testing, design of experiments, quality control, multivariate
data reduction and robustness, among others. Many of the
current major advances in statistics rely heavily on sophis-
ticated mathematics.

And yet, the formal structures of mathematical statistics
seem to have a blind spot. Most of the mathematical development has to do with pre-data analysis: Is such-and-such
likely to be a good procedure? How should we plan to do
so-and-so? To answer such questions requires one to embed
one's particular statistical problem a priori in a sample
space with superimposed probability distribution over

Scientific Inference,
Data Analysis, and Robustness

potential realizations. The blind spot is the implicit
assumption that pre-data probability calculations are relevant to post-data inferences. This blind spot is covered by
Bayes's Theorem, which explicitly recognises the difference
between pre-data and post-data contexts. Must the blind-
spot remain in frequentist statistics? First we must recognize that it exists, and that is one purpose of the present
paper: to show by example how naive frequentist answers can
be misleading.

Some of our difficulties can be traced to the way in
which we learn and teach probability and statistics. While
we place great emphasis on the calculus of probability, we
say little about the practical use of probability—other
than to endorse the simple relative frequency interpretation.
The following three probability statements suggest a variety
of application that we would do well to understand and teach

  P(coin will land head-up on a single flip) = ½

  P(rain tomorrow in Madison) = ½

  P(nuclear war in Europe before 1990) = ½.

It is evident that relative frequency in a real sequence of
repeated experiments is simply not a rich enough interpretation of probability.

When we move from probability to theoretical statistics,
we start by introducing such problems as "Let X_1, ..., X_n
be i.i.d. with p.d.f. $f_\theta(x)$..." and "Let $Y_i = \beta_0 + \beta_1 X_i + \epsilon_i$,
where ϵ_1, ..., ϵ_n are i.i.d. $N(0, \sigma^2)$." What meanings might
these statements have? How do we determine that such state-
ments are reasonable, even as approximations? I believe
that there is far too little integration of inference with
modelling and model diagnosis. For example, where do we read
about inference subject to adequacy of model fit as judged by
a goodness-of-fit test?

Once exact models are established, theoretical discussion
focusses on exactness: unbiasedness, sufficiency, locally
most powerful, inadmissibility, ancillarity, and so on. Of
course these exact concepts are useful, in part because they
prevent loose thought and ad hocery—but the concepts and
exact properties should be used only as guides for careful

approximate analysis. Pedantic obsession with exactness can lead to absurdity, as a simple example will illustrate.

Example 1. Suppose that a population of N elements, with associated measurements $(\tilde{x}_1, \tilde{y}_1), \ldots, (\tilde{x}_N, \tilde{y}_N)$, is sampled randomly n times with replacement. Let the generic sampled values be denoted by $X_i = \tilde{x}_{I_i}$ and $Y_i = \tilde{y}_{I_i}$, $i = 1, \ldots, n$; of course $\Pr(I_i = k) \equiv N^{-1}$. A graph of the data strongly suggests that $\tilde{y} \doteq \beta\tilde{x}$, and deviations from this approximate relationship seem reasonably random except for absolute magnitude. One might then propose the working model

$$Y_i = \beta X_i + \varepsilon_i \quad ,$$

with ε_i independent such that $E(\varepsilon_i | X_i = x) = 0$ and $\text{Var}(\varepsilon_i | X_i = x) = \sigma^2(x)$, where the form of $\sigma^2(\cdot)$ is suggested by the sample. But now suppose that $\tilde{x}_1, \ldots, \tilde{x}_N$ are known and that they are distinct (not an unusual occurence). Then it cannot be true that $E(\varepsilon_i | X_i = x) = 0$, since $E(\varepsilon_i | X_i = \tilde{x}_j)$ $= \tilde{y}_j - \beta\tilde{x}_j$, and further $\text{Var}(\varepsilon_i | X_i = \tilde{x}_j) \equiv 0$. This truth is useless to practical analysis of the data. One reasonable mathematical way to proceed is by acknowledging that "$E(\varepsilon_i | X_i = x) = 0$" stands for "$|\text{ave}(\tilde{y}_i - \beta\tilde{x}_i | x - \delta \leq \tilde{x}_i \leq x + \delta)| < \varepsilon$" for suitably small δ and ε—if a full-blown mathematical description is needed, which is doubtful.

This brief general discussion has raised questions about the difference between pre- and post-data probability calculations, the legitimacy of exact theory when integrated with practical application, and careful specification and understanding of what a statistical model means in practical terms. The purpose of the more detailed discussion which follows is to consider four topics where naive application of frequentist statistical theory can lead to incorrect or unhelpful inferences, whereas careful attention to the above questions can lead to sensible frquentist inferences.

The four topics to be discussed are: likelihood inference, inference from transformed data, randomization in design of experiments and surveys, and robust estimation.

One main thrust of the paper is that the general concept of ancillarity is an integral part of statistics.
Beyond that, one might conclude that Bayesian inference provides a simple, direct way of obtaining sensible answers.
This does not conflict with the view that frequency provides
operational substance to many inferences. It is my curious
belief that careful frequentist and careful Bayesian approaches can complement one another.

2. Likelihood Inference

The most clearly developed framework for inference
principles and methods is that of stochastic models with a
single unknown parameter. A conventional problem would
treat observations $(x_1, \ldots, x_n) = \underset{\sim}{x}$ as a realization of random
variables $(X_1, \ldots, X_n) = \underset{\sim}{X}$ whose joint distribution has den-
sity $f(\underset{\sim}{x}|\theta)$. Our discussion begins by reviewing what is
known for the case of independent X_i, explained for simpli-
city under the assumption of homogeneity--that is, $f(\underset{\sim}{x}|\theta) =$
$\Pi\ g(x_j|\theta)$. Part of our purpose is to expose an alternative
large-sample approximate theory for likelihood inference,
and to suggest by example how widely applicable that approximate theory may be.

There is agreement that for inference under a given
model assumption, $\underset{\sim}{x}$ should first be reduced to the minimal
sufficient statistic. In general this does not simplify
matters much, so we must deal with approximate sufficiency.
One program is to reduce

$$\underset{\sim}{X} \rightarrow (\hat{\theta}, A_1, \ldots, A_p) = (\hat{\theta}, \underset{\sim}{A})$$

where $\hat{\theta}$ is the MLE, and $\underset{\sim}{A}$ contains transformed characteristics of the likelihood shape. The larger is p, the less
information is lost in the reduction in general. In particular, if p = 1 and if A_1 is the studentized form of the ob-
served information $I = -\{d^2 \log f(\underset{\sim}{x}|\theta)/d\theta^2\}_{\theta=\hat{\theta}}$, then the
information lost in reducing $\underset{\sim}{x} \rightarrow (\hat{\theta}, I) \equiv (\hat{\theta}, A_1)$ is o(1) as
$n \rightarrow \infty$. (Reduction $\underset{\sim}{x} \rightarrow \hat{\theta}$ does not have this property.) It
is fairly clear that the rest of the likelihood shape, for-
malized in $(A_2, \ldots)$, can be ignored if the likelihood is
very nearly normal in shape, which is often the case for

quite moderate sample size. It is also very clear from examples that the variation in spread of normal-shaped likelihoods can be appreciable, which is why reduction $x \to \hat{\theta}$ is
often inadequate--as calculation of s.e. (I) would indicate.

Given appropriate (non-unique) definitions of $A_1,\ldots A_p$,
one can show that

(i) $I^{\frac{1}{2}}(\hat{\theta}-\theta) \approx N(0,1)$ given $\underset{\sim}{A}=\underset{\sim}{a}$,   (2.1)

(ii) $I/E(I) \approx N(1,n^{-1}c_\theta^2)$ $c_\theta = O(1)$,

(iii) $\underset{\sim}{A} \approx N_p(0,1)$.

(For simplicity I write E(I) in place of $E\{-d^2 \log f(x|\theta)/$
$d\theta^2\}_{\theta=\hat{\theta}}$, which it approximately equals.) These distributional approximations form the basis for an <u>approximate</u> an-
alysis based on the <u>approximate</u> sufficient statistic $(\hat{\theta},\underset{\sim}{A})$,
and of course more refined approximations may sometimes be
needed. If we were to reduce $x \to \hat{\theta}$, (i)-(iii) would not be
directly relevant, but (i) and (ii) do imply

(i') $\{E(I)\}^{\frac{1}{2}} (\hat{\theta}-\theta) \approx N(0,1)$ (2.2)

unconditionally, which is the classical first approximation
result for the large-sample distribution of $\hat{\theta}$. The unusual
derivation of (i') from (i) and (ii) will be important below.

The expressions (i) and (iii) give the (approximate)
sufficient pivotal inference "$I^{\frac{1}{2}}(\hat{\theta}-\theta)$ is a standard normal
random variable"--statement of the value $\underset{\sim}{a}$ of $\underset{\sim}{A}$ is uninformative. Why use the conditional statement (i) rather than (i')?
To answer this in detail would be to repeat well-worn arguments for the conditionality principle, which we need not do;
see Efron & Hinkley (1978). One brief answer will be given:
if inference is contingent on adequacy of model fit, then the
same (conditional) pivotal inference statement is valid, because the fit of the model is tested using $\underset{\sim}{A}$ (for example,
the chi-squared statistic ΣA_j^2)--therefore the model adequacy
restriction is covered by the conditioning on $\underset{\sim}{A}=\underset{\sim}{a}$. By contrast, the distributional approximation (i') may be recog-
nizably invalid for subsets of the <u>a priori</u> sample space which
are determined by restrictions on $\underset{\sim}{A}$.

It should be clear that (2.1) (i) may not be sufficiently accurate. Both Cox (1980) and Hinkley (1980) show that the likelihood itself can be integrated to give a more accurate approximation for the conditional distribution of $I^{\frac{1}{2}}(\hat{\theta}-\theta)$. This brings us tantalizingly close to the formal Bayesian posterior distribution. Notice that the Bayesian analysis requires no choice of pivot, which might prompt a frequentist to investigate the whole class of approximate conditional inferences based on pivots $Q(\hat{\theta},\theta,A_1,\ldots,A_p)$, rather than restricting attention to $I^{\frac{1}{2}}(\hat{\theta}-\theta)$. Of course the search for exact equivalence of conditional frequentist and Bayesian inferences is in principle hopeless, but it is of interest to study the approximate formal equivalence, since this gives added force to the Bayesian method.

What has been outlined here is a means of approximate analysis based on the actual information content of the data. The reader should consult Barndorff-Nielsen (1980), Amari (1982a,b), and other cited references, for theoretical details.

The question I should like to address now is the extent to which our discussion generalizes, beyond the case of independent sampling with fixed sample size. This draws us back to the relationships (2.1) (i) and (ii). No general theoretical account seems to be possible at this stage, so we shall look at two interesting examples. The first deals with a non-stationary process, and the second deals with random sample size.

Example 2. Autoregressive Process

Let $X_1,X_2,\ldots,X_n$ be modelled by the AR1 process

$$X_o = 0, \qquad X_j = \theta X_{j-1} + \varepsilon_j , \qquad j=1,\ldots,n ,$$

where $\varepsilon_1,\varepsilon_2,\ldots$ are independent $N(0,1)$. The earlier discussion applies if $|\theta| < 1$, that is to say (2.1) and (2.2) hold, for large n, essentially because the process is ergodic. But for $|\theta| \geq 1$, when the process is not ergodic, it is well known that (i') fails. What seems not to be well known is

that (i) still holds for $|\theta| \geq 1$, but (ii) fails. This re-
inforces the idea that (i) is the fundamental normal
approximation of likelihood inference.

The loglikelihood for θ based on $x_1, \ldots, x_n$ is

$$\ell(\theta) = \text{constant} + \theta \sum_1^n x_j x_{j-1} - \tfrac{1}{2}\theta^2 \sum_1^n x_{j-1}^2 ,$$

so that the sufficient statistic is the pair $(\hat{\theta}, I)$ where

$$\hat{\theta} = \sum_1^n x_j x_{j-1} / \sum_1^n x_{j-1}^2 \quad \text{and} \quad I = -\ddot{\ell}(\hat{\theta}) = \sum_1^n x_{j-1}^2 ;$$

the normalized likelihood is exactly $N(\hat{\theta}, I^{-1})$. For $|\theta| \geq 1$,
the magnitude of variation in I prevents convergence of
$I/E(I)$, and (i') fails: another unconditional limiting dis-
tribution obtains for $\{E(I)\}^{\frac{1}{2}} (\hat{\theta}-\theta)$. The insufficiency of
$\hat{\theta}$ alone, and the huge asymptotic fluctuations of I, make the
unconditional limiting result clearly irrelevant for finite
sample analysis. The sufficient pivotal result $I^{\frac{1}{2}}(\hat{\theta}-\theta) \approx$
$N(0,1)$ has apparently been established for $\theta=1$ and large n
by both D. Siegmund and T. Lai, and empirical evidence such
as is described below suggests validity of the result for
$|\theta| \geq 1$.

The unconditional theory for the case $\theta=1$ has been
studied by Evans and Savin (1981), who enumerate the non-
normal limiting distribution of $\{E(I)\}^{\frac{1}{2}}(\hat{\theta}-\theta)$; here $E(I) =$
$\tfrac{1}{2}n(n-1)$ is the true expectation of I at $\theta=1$. Figure 1 shows
(dotted curve) the induced approximate unconditional distribution of $\hat{\theta}-\theta$ for n=20. The figure also shows (dashed lines)
the induced conditional normal approximations for $\hat{\theta}-\theta$ ob-
tained by "undoing" $I^{\frac{1}{2}}(\hat{\theta}-\theta) \approx N(0,1)$ when $I=\tfrac{1}{4}$, 1 and 2 times
$E(I)$. (Simulation shows that when n=20, $\Pr\{I \leq \tfrac{1}{4}E(I)\} \approx \tfrac{1}{4}$
and $\Pr\{I \geq 2E(I)\} \approx 1/8$, so that Figure 1 encompasses a
reasonable span for I.)

Evidently inference from the unconditional approximate
distribution would be very misleading even if $I = E(I)$. For
example, still with n=20, the unconditional result indicates
that values of $\hat{\theta}$ smaller than $\theta - 0.4 = 0.6$ would be significant evidence that θ is less than 1, rather than equal to

Figure 1: Comparison of induced distributions for $\hat{\theta}-\theta$
in AR1 process with $x_0=0$ and $n=20$. Dotted line:
unconditional asymptotic distribution when $\theta=1$.
Dashed line: conditional normal approximations induced
from $N(0,1)$ approximation for $I^{\frac{1}{2}}(\hat{\theta}-\theta)$.

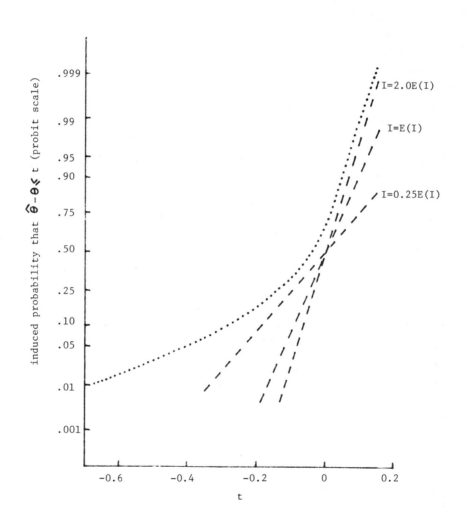

Figure 2: Rankit plots of 200 simulated values of pivot
$P=I^{\frac{1}{2}}(\hat{\theta}-\theta)$ for AR1 process with $x_O=0$ and $n=20$.
Upper graph: $\theta=1$. Lower graph: $\theta=2$.

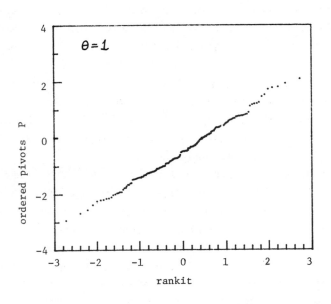

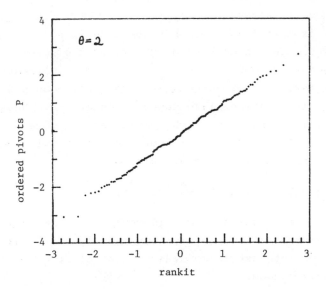

one, at the 5% level. But the conditional significance level
at $\hat{\theta}$ = 0.6 is 5% for I = 1/10E(I), less than 1% for I = ¼E(I)
and essentially zero for I $\geq$ E(I).

Should econometricians test θ=1 using the unconditional
result, there will be a lot of random walks in econometric
models!

Are the approximations in fact of reasonable accuracy?
Yes and no. Figure 2 shows rankit plots of 200 simulated
values of $I^{\frac{1}{2}}(\hat{\theta}-\theta)$ for the cases θ=1.0 and 2.0 with x_o=0 and
n=20. The plots for θ near 1 show a perceptible bias, which
disappears for θ=2. The folded distributions agree very well
with theoretical approximation: in 10,000 cases with θ=1 the
frequencies with which $I(\hat{\theta}-\theta)^2$ exceeded the 2%, 5% and 10%
points of χ_1^2 were respectively 1.98%, 4.99% and 9.98%.

Example 3.   Sequential Sampling

Let $\{X(t):t \geq 0\}$ be a Wiener process with drift rate θ,
i.e. increments dX(t) are independent N(θdt,dt). Anscombe
(1957) considered this model under two sampling rules: (a)
stop sampling at T=t_o, (b) stop sampling when first X(t)=x_o.
In general, the stopping time T and $X \equiv X(T)$ are jointly sufficient for data-dependent stopping rules, and so a possibility of approximate ancillarity exists. But $\hat{\theta}$=X/T is sufficient under either of rules (a) and (b). The normalized
likelihood is always exactly N($\hat{\theta}$,I^{-1}), $I \equiv T$. Under rule (a),
$I^{\frac{1}{2}}(\hat{\theta}-\theta)$ is exactly N(0,1), but the same is not so under rule
(b). Thus, as Anscombe pointed out, at the level of exact
inference it is not always possible to obtain frequency-
based inference that agrees with the likelihood principle--
a principle fundamental to Bayesian inference. Nevertheless,
in the sample space in which samples are embedded to obtain
frequency, $I^{\frac{1}{2}}(\hat{\theta}-\theta)$ is still approximately N(0,1). The exact
density of $Q = I^{\frac{1}{2}}(\hat{\theta}-\theta)$ under rule (b) is

$$p(q|x_o,\theta) = \{1 + \tfrac{1}{2}(\theta-\hat{\theta})\}^{-1} \phi(q) \quad ,$$

where $\hat{\theta}$ is a function of q.   (Somewhat mischievously, the
usable approximation $p(q|x_o,\hat{\theta})$ is exactly $\phi(q)$!)   It would
be of some interest to consider a general stopping rule with
(T,X) both random.

A rather different sequential scheme was investigated by Grambsch (1980). She showed that for independent $X_1, X_2,$ $\ldots, X_N$ with $N = \inf(n : I \geq i_o)$, still $I^{\frac{1}{2}}(\hat{\theta}-\theta) \approx N(0,1)$.

What these two examples suggest is that "$I^{\frac{1}{2}}(\hat{\theta}-\theta) \approx N(0,1)$" is a very general normal approximation, appropriate when the likelihood is approximately normal shaped. Or, put another way, the result of applying Bayes's theorem has approximate validity in a frequency sense if the sample sequence is embedded in a suitably small subset of the sample space--this "suitably small subset" being determined by an approximately ancillary statistic when possible. Much further work is needed before these vague ideas and suggestions can be turned into a concrete, reliable approximate theory--if this is indeed possible. What is not in doubt, however, is that unconditional asymptotic theory for $\hat{\theta}$ is insufficient both in the technical sense and in the practical sense.

3. Data-based Transformation: Box-Cox Revisited Again

A conventional mathematical formulation for linear regression on explanatory variables $\underset{\sim}{u}$ with random deviations is

$$Y_i = \theta^T u_i + \sigma \varepsilon_i \quad , \quad i=1,\ldots,n \quad , \tag{3.1}$$

where $\varepsilon_1, \ldots, \varepsilon_n$ are independent $N(0,1)$. In practise Y may be a derived measurement, e.g. logarithm or reciprocal of an actual measurement, and clearly some empirical judgement should be involved in such a particular choice of Y. Box & Cox (1964) proposed and studied a particularly useful class of derived measurements obtained from actual measurements X, namely $Y = (X^\lambda-1)/\lambda$. The data then provide evidence about those values of λ which are compatible with models of the type (3.1). The literal extension of (3.1) is

$$X_i(\lambda) = (X_i^\lambda-1)/\lambda = \theta^T u_i + \sigma \varepsilon_i \quad , \quad i=1,\ldots,n \quad . \tag{3.2}$$

A maximum likelihood approach to model fitting is first to obtain $\hat{\lambda}$, and then to apply standard least squares to data $\{(u_i, x_i(\hat{\lambda})), \quad i=1,\ldots,n\}$. But what of inference?

Bickel & Doksum (1981) show that under model (3.2), the distributions of $\hat{\theta}$ and $\hat{\sigma}^2$ can be very much more dispersed than
if λ were known, and hence not estimated. For example, if
$u \equiv 1$ and $\lambda = 0$, then $\sqrt{n}(\hat{\theta}-\theta) \approx N(0,\sigma^2+\zeta^2)$, $\zeta^2 = 1/6\sigma^2$
$(1+\theta^4/\sigma^4)$--the extra term ζ^2 being "due to estimation of λ".
To my chagrin, this particular result was first noted by
myself (Hinkley, 1975).

Bickel & Doksum's remarkable observation is mathematically correct, but statistically incorrect: the variance in-
flation is not relevant to data analysis. This is an instance where the mathematical model has come adrift from its
motivating moorings. A simple example will clarify the main
point.

Suppose that we have two samples of positive measurements x--so that associated u vectors are (1,1) and (1,-1),
say--and that we wish to compare the two sampled populations.
The simplest type of comparison is via means, so we fit
model (3.2) and find $\hat{\lambda}=0$. In all respects the two samples
appear to suggest that (3.2) fits well. Whatever be the
true population λ, our data model says that $\log_e X \sim N(\mu_i, \sigma^2)$
in sample i, i=1,2. For reasonably large n, then, our com-
parison is summarized by the following statement

$$"(\mu_1-\mu_2) - (\hat{\mu}_1-\hat{\mu}_2) \approx N(0,\sigma^2(1/n_1+1/n_2)) \tag{3.3}$$

where μ_i is the mean of $\log_e X$ in population i"; n_1 and n_2
are sample sizes. In contrast to this statement is one based
on Bickel & Doksum's results, which takes the form

$$"\theta_2-\hat{\theta}_2 \approx N(0,\sigma^2(1/n_1+1/n_2) + \zeta^2(1/n_1+1/n_2)) \tag{3.4}$$

where θ_2 is ...". θ_2 is what? All that can be said is that
θ_2 is the mean population contrast of some unknown transformation of X! The point is that (3.3) is a scientifically
complete statement, whereas (3.4) is not. Further, (3.3)
matches the interplay between data modelling and statistical
inference--it is the same statement that would be made if we
arrived at the log transformation by some other logical exploratory method.

The statement (3.3) deserves an extended discussion, which will be given elsewhere (Hinkley and Runger, 1982). Here only the key points will be mentioned. First, (3.3) is the inference in a context which does not make $\mu = \log_e X$ of a prior interest--that would involve a different analysis, clearly, since $\log_e X$ need not always appear to be normally distributed. Secondly, the probability result used in (3.3) needs justification. In the absence of the extended discussion, I will note that (3.3) is precisely the statement that you, the reader, would use if I presented you with the Normal-looking values of $\log_e X$ and asked for a statement about $\mu_1 - \mu_2$. Thirdly, the difficulty can be ascribed to incompleteness of formulation: (3.2) should really be

$$X_i(\lambda) = \{\theta(\lambda)\}^T u_i + \sigma(\lambda)\varepsilon_i \quad , \quad i=1,\ldots,n \quad ,$$

where roughly speaking $\theta(m) = E(\hat{\theta}|\hat{\lambda}=m)$. (Box and Cox rescale $x(\lambda)$ to offset the incomparability of $\theta(\lambda)$ values as λ varies.) The Bickel & Doksum phenomenon of variance inflation is simply that

$$\text{Var}(\hat{\theta}) = E \text{ Var}\{\hat{\theta}(\hat{\lambda})|\hat{\lambda}\} + \text{Var } E\{\hat{\theta}(\hat{\lambda})|\hat{\lambda}\}$$

where the second term on the right is the inflation--which we necessarily discount when we fix our scale by $\hat{\lambda}_{obs}$ and make an inference about $\theta(\hat{\lambda}_{obs})$. Fourthly, there is the Bayesian approach. Note that the marginal posterior for θ will give, approximately, (3.4) and this is deemed to be irrelevant. What is required, if $\hat{\lambda}=0$ as before, is the posterior distribution for $\theta_2(0) = \mu_1 - \mu_2$. This distribution can be derived, approximately, by writing $\theta(0) \approx \theta(\lambda) - \lambda \dot{\theta}_2(0)$. I leave the reader to follow this approach through to (3.3).

I have already mentioned that for a parameter of prior interest, such as $\mu_1 - \mu_2$, the analysis would be different. So it would for a predictive statement about the observable X. For inference relative to a fixed <u>a priori</u> scale, the Bickel & Doksum results do not apply mathematically or statistically. For some relevant work see Carroll and Ruppert (1981).

The lessons to be learned here are (i) that care must
be exercised in model statement (compare (3.2) and (3.3)),
and (ii) that inference statements must be physically complete, i.e. well defined and relevant.

4. Randomized Designs and Their Analyses

Two subjects are of interest here, classical experimental design and sample survey analysis, the common element
being randomization--in the first case used to allocate
units to treatments, and in the second case used to select
units for measurement. My main point is that the general
concept of ancillarity plays an important role in both design
and analysis, in the sense of pre- and post-design blocking.

Symbolic summaries of the randomized design problems
would be as follows:

Experimental Design: Given a set of designs $\mathcal{D} = \{d_1, \ldots, d_M\}$,
apply the random design D where $\Pr(D=d_j) \equiv M^{-1}$.

Sample Survey (with Replacement): Given population units
$\tilde{u}_1, \ldots, \tilde{u}_N$, a design d is a point in $\{\tilde{u}_1, \ldots, \tilde{u}_N\}^n$. From the
set $\mathcal{D} = (d_1, \ldots, d_M)$, $M=N^n$, choose a random design D such
that $\Pr(D=d_i) \equiv M^{-1}$. (D involves individual selections
$U_1, \ldots, U_n$ such that $\Pr(U_i=u_j) \equiv N^{-1}$.)

It is sometimes argued that D is an ancillary statistic
and hence that randomization theory of statistics is incom-
patible with the theory of conditioning on ancillary statistics. This interesting twist of logic belies the purpose
of an ancillary statistic, which is to act as an indicator
of appropriate reference sets. (Part of the absurdity of
the exact logic was mentioned in Example 1.) One point that
has been made is that randomization is often used to make
normal-theory analysis valid (approximately), in which case
D is ancillary by design. A second point is that the ancillary statistic should partition $\mathcal{D}$ only as far as is statistically useful--i.e. as far as the relevant subsets of
the a priori sample space. Two examples will clarify the
situation.

Example 4. Knight's Move Latin Squares

In 1931, Tedin reported on a detailed analysis of the
5x5 Latin Squares as applied to some uniformity data (agricultural plot responses in the absence of treatment). These

particular data evidenced spatial correlation. The 5x5
Latin Squares $\mathfrak{D}$ can be subdivided into $\mathfrak{D}_1$ = Knight's Move
Squares, $\mathfrak{D}_2$ = Diagonal Squares, $\mathfrak{D}_3$ = Other Squares. If we
denote by σ_i the randomization standard error of a treatment
contrast conditional on $D\epsilon\mathfrak{D}_i$, then Tedin showed that $\sigma_1 < \sigma_3$
$< \sigma_2$ for his uniformity data. However, the estimated vari-
ance $\hat{\sigma}^2$ based on residual sum of squares behaves in contrary
fashion, the restricted randomization means satisfying

$$E(\hat{\sigma}^2 | D\epsilon\mathfrak{D}_1) > E(\hat{\sigma}^2 | D\epsilon\mathfrak{D}_3) = \sigma_3^2 > E(\hat{\sigma}^2 | D\epsilon\mathfrak{D}_2) \ .$$

The central equality validates normal-theory estimation of
standard error for the restricted randomization over $\mathfrak{D}_3$.
Thus the ancillary indicator of design subset can be used in
conjunction with empirical data to facilitate design.

Of course if reliable estimates of σ_i were available,
one could select D from $\mathfrak{D}$ without restriction and then use
the ancillary subset indicator in the data analysis to obtain
the relevant standard error of a contrast; $\mathfrak{D}_1$ would be a
preferable restriction in such a situation. If a spatial
correlation model accounted for the differences noted by
Tedin, then presumably appropriate analyses under that model
would distinguish between $\mathfrak{D}_1$, $\mathfrak{D}_2$ and $\mathfrak{D}_3$ on the basis of an
(approximately) ancillary statistic.

Further discussion of this example may be found in
Yates (1965).

Example 5. Ratio Estimation of a Population Total

The following situation seems to be common in some cen-
sus problems. For a population of units $(\tilde{u}_1, \tilde{u}_2, \ldots, \tilde{u}_N)$,
each of which possesses quantitative characteristics x and
y, there is complete knowledge of $\tilde{x}_1, \ldots, \tilde{x}_N$. It is now
desired to estimate $T = \sum_1^N \tilde{y}_i$, and the pairs (x,y) can be measured on a random sample of units. Thus we obtain a random
sample $\{(x_1, y_1), \ldots, (x_n, y_n)\}$. The ratio estimator of T is

$$\hat{T} = (\Sigma_1^N \tilde{x}_i)\bar{y}/\bar{x} \ , \qquad \bar{x} = n^{-1}\Sigma_1^n x_j \ , \qquad \bar{y} = n^{-1}\Sigma_1^n y_j \ .$$

Several estimates V of $Var(\hat{T})$ have been proposed, and
some are justified on the grounds that their means over the
full randomization distribution are nearly unbiased. Such
justification is rightly challenged by Royall and Cumberland
(1981), on the well-documented empirical evidence that often
the values of $x_1,\ldots,x_n$ define relevant subsets of the full a
priori sample space. Thus $Var(\hat{T}|x_1,\ldots,x_n)$ and $E(V|x_1,\ldots,$
$x_n)$--carefully interpreted--may be very unequal, as they both
vary with the x_i. This could often be anticipated because in
many cases $Y \approx \beta X + \sigma(x)e$ is a plausible model. Then the
relationship between $(x_1,\ldots,x_n)$ and $(\tilde{x}_1,\ldots,\tilde{x}_N)$ can be used
as an ancillary indicator in a conditional inference, based
on estimate $\hat{T}$. In effect the indicator would act as a post-
data stratification instrument.

In both examples it is suggested that a statistical indicator a(D) be used to split D into subsets, either for the
purpose of design or for the purpose of analysis. In the
latter case, the value of a(D) will define a subset on which
the inferential probability is defined, either by restricted
randomization or by some plausible model that is validated
by the randomization.

5. Robust Estimation

A great deal of effort has gone into the following problem: if $x_1,\ldots,x_n$ are independently distributed each with
density of the form $f(x-\theta)$ where $f(y)$ is symmetric about $y=0$,
what is a good estimate for θ if the form of f is unknown?
Our language has been enriched by terms such as contamination,
leverage and break-down point, and dozens of estimators have
been studied intensively--both by brains and by computers.
One thing seems to have been overlooked: how does one analyse
a given data set?

A common approach to robustness theory might be de-
scribed simply as follows: Consider a class $\mathcal{T}$ of statistical
functionals $t(\cdot)$ such that $t(\hat{F}_n)$ is an unbiased estimate of
θ, $\hat{F}_n$ being the empirical distribution function. Then if
attention can be restricted to a class $\mathcal{F}$ of underlying

distribution functions F, choose the estimate $t(\hat{F}_n)$ to obtain $\min_{t \in \mathcal{T}} \max_{F \in \mathcal{F}} E\{t(F_n)-\theta\}^2$. Often $\mathcal{F}$ would be a neighborhood of a special distribution, such as the Normal.

An asymptotic theory for statistical functionals gives unconditional normal approximations for $t(\hat{F}_n)-\theta$. But this seems unsatisfactory on several grounds. First, parametric inference shows convincingly (Efron & Hinkley, 1978) that the relevant precision for $t(\hat{F}_n)$ should be tied to the residuals $\hat{e}_i = x_i - t(\hat{F}_n)$. Secondly, statisticians diagnose data with the aid of things such as normal plots, and can often gain information about which F is appropriate by using the residuals. Should we worry that F might be capable of producing very large deviations x-θ, if in fact the given data yield the best normal plot we ever saw? Clearly not. Related to this is the more specific conjecture: that whatever F is, $\overline{X}$ is a pretty good estimate if a goodness-of-fit test of normality does not reject the normality hypothesis. This seems plausible because for any choice of t, $\overline{X} = t(\hat{F}_n) + d_t(\hat{e})$, whereas a goodness-of-fit statistic for a particular F must be of the form $g_{t,F}(\hat{e})$. The covariation of $(d_t(\hat{e}), g_{t,F}(\hat{e}))$ will keep $d_t(\hat{e})$ in check if $g_{t,F}(\hat{e})$ lies near its expected value under F.

Would it be reasonable to choose the estimator t and calculate its standard error as if $F=F_0$ when the observed data are compatible with F_0? In a general sense one would say "yes", because this is how applied statisticians behave when they first model their data. The theoretician would be doubtful. The key question might be posed in the following simple form: Suppose $\mathcal{F} = (F_1, \ldots, F_m)$, that

$$n \text{Var}\{t(\hat{F}_n)|F_i, \hat{e}\} \approx \sigma_i^2(\hat{e}) \quad ,$$

and that $\hat{F}_{\mathcal{F}}$ is the F_j which is closest to $\hat{F}_n$ (in an appropriate sense). Is it then true that

$$n \text{Var}\{t(\hat{F}_n)|F_i, \hat{F}_{\mathcal{F}}=F_k, \hat{e}\} \approx \sigma_k^2(\hat{e})?$$

(The same question is of interest for unconditional variances.) If true, this would parallel a Bayesian analysis
for relatively uninformative priors:

$$p(\theta|x_1,\ldots,x_n)=C\sum_{F\epsilon\mathcal{F}}p(\theta|x_1,\ldots,x_n,F)p(F|x_1,\ldots,x_n)p_{PRIOR}(F)$$

$$\approx p(\theta|t(\hat{F}_n),\hat{e},\hat{F}_{\mathcal{F}}),$$

which is approximately a $N(0,n^{-1}\sigma_k^2(\hat{e}))$ distribution for
$\theta-t(\hat{F}_n)$ when $\hat{F}_{\mathcal{F}}=\hat{F}_k$.

Of course a Bayesian analysis has the general advantage
of responding to specific features of the data. What I am
suggesting is that careful attention to robust inference
might reveal a sensible, responsive frequency theory which
essentially justifies the results of a Bayesian analysis.
The unconditional distribution theory prevalent in robustness
literature is fine for choosing estimates which are generally
good, but not for analysis of a particular data set.

REFERENCES

AMARI, S. (1982a). Differential geometry of curved exponen-
    tial families--curvatures and information loss. Ann.
    Statist., 10 (June issue).
AMARI, S. (1982b). Geometrical theory of asymptotic ancil-
    larity and conditional inference. Biometrika, 69,
    1-17.
ANSCOMBE, F.J. (1957). Dependence of the fiducial argument
    on the sampling rule. Biometrika, 44, 464-469.
BARNDORFF-NIELSEN, D. (1980). Conditionality resolutions.
    Biometrika, 67, 293-310.
BICKEL, P.J. and DOKSUM, K.A. (1981). An analysis of trans-
    formations revisited. J.Am.Statist.Assoc., 76, 296-311.
BOX, G.E.P. and COX, D.R. (1964). An analysis of transforma-
    tions (with discussion). J.R.Statist.Soc., B26, 211-252.
CARROLL, R.J. and RUPPERT, D. (1981). On prediction and the
    power transformation family. Biometrika, 68, 609-615.
COX, D.R. (1980). Local ancillarity. Biometrika, 67, 279-
    286.

EFRON, B. and HINKLEY, D.V. (1978). Assessing the accuracy
    of the maximum likelihood estimator: Observed versus
    expected Fisher information. Biometrika, 65, 457-482.

EVANS, G.B.A. and SAVIN, N.E. (1981). The calculation of the
    limiting distribution of the least squares estimator of
    the parameter in a random walk model. Ann.Statist., 9,
    1114-1118.

GRAMBSCH, P. (1980). Likelihood inference. Ph.D. Disserta-
    tion, University of Minnesota.

HINKLEY, D.V. (1975). On power transformations to symmetry.
    Biometrika, 62, 101-111.

HINKLEY, D.V. (1980). Likelihood as approximate pivotal dis-
    tribution. Biometrika, 67, 287-292.

HINKLEY, D.V. and RUNGER, G. (1982). Analysis of Box-Cox
    transformed data. University of Minnesota School of
    Statistics Technical Report.

ROYALL, R.M. and CUMBERLAND, W.G. (1981). An empirical study
    of the ratio estimator and estimators of its variance
    (with discussion). J.Am.Statist.Assoc., 76, 66-88.

TEDIN, O. (1931). The influence of systematic plot arrange-
    ment upon the estimate of error in field experiments.
    J.Agric.Sci.Camb., 21, 191-208.

YATES, F. (1965). A fresh look at the basic principles of
    the design and analysis of experiments. Proc. Fifth
    Berkeley Symposium on Mathematical Statistics and Prob-
    ability, 4, 777-790.

The author was partially supported by National Science
Foundation Grant MCS-79-04558.

School of Statistics
University of Minnesota
Minneapolis, MN 55455

Frequency Properties
of Bayes Rules

Persi Diaconis and David Freedman

ABSTRACT

The "what if" method uses frequentist computations as a
way of thinking about prior distributions. Several applications to priors on infinite dimensional parameter spaces
give backing to the widely held belief that it is hard to
understand priors on large spaces. Examples include
Dirichlet priors, and priors on the continuous functions.
In both settings, new examples of inconsistent Bayes rules
are given.

1. The "What if" Method

It is clear that prior distributions are useful tools
for frequentist analysis. Indeed, well known complete class
theorems say that any admissible procedure is approximately
Bayes. We want to discuss a rationale for a Bayesian to be
interested in frequency calculations like consistency and
robustness. The idea is simple: After specifying a prior
distribution; generate imaginary data sequences; compute the
posterior distribution; and consider if the posterior is an
adequate reflection of current knowledge. The exercise is
proposed as a method of thinking about prior distributions.
It is quite close to the checks for coherence proposed by
de Finetti and Savage.

Scientific Inference,
Data Analysis, and Robustness

105

Consider a simple example: One of us was interested in predicting the outcome of a classroom demonstration involving 50 spins of a coin on its edge. Prior experience suggests that when coins are spun on edge, as opposed to flipped in the air, they are very far from fair (try it!). The bias can be 70% or more, some coins being biased towards heads, some towards tails. The knowledge just described can be incorporated as a bimodal prior on [0, 1] say

$$p \sim \frac{1}{2} \beta(10, 20) + \frac{1}{2} \beta(20, 10),$$

with $\beta(a, b)$ denoting a beta distribution with parameters a and b. Having gone this far, some reflection suggested that in the past more coins were biased towards tails. As a further check, the following mental exercise was conducted: "What if in 50 spins, about half came out heads and about half came out tails?" This possiblity did not seem so unreasonable, after all some coins are probably fairly well balanced. The final prior used was

$$p \sim \frac{1}{2} \beta(10, 20) + \frac{1}{10} \beta(15, 15) + \frac{2}{5} \beta(20, 10).$$

The use of fictitious samples was suggested by Good's "device of imaginary results". Good [1950, p. 35] uses the device as a method of roughly quantifying a prior in difficult situations. We regard the method as useful in thinking about priors: "What if the data turned out like ... ." Hence the name.

There has been a good deal of recent work on priors in infinite dimensions. It is widely acknowledged that it is hard to understand priors in high dimensional problems. See the discussion in Efron (1970). The following example is meant to illustrate how far off our intuition can be in a thoroughly studied situation: The Dirichlet prior over the set of cumulative distribution functions on the line. A convenient description of this prior is in Ferguson (1974). We here recall that the prior is based on a positive measure α. If $A_1, \ldots, A_k$ is a partition of the real line, the random variable $[P(A_1), \ldots, P(A_k)]$ is to have a Dirichlet distribution with parameter $\alpha(A_1), \ldots, \alpha(A_k)$. As Ferguson shows, this prior has some appealing properties. For example, the Bayes estimate of the underlying measure given a sample of size n is a convex combination of the empirical distribution function and the measure α normalized as a probability. Still, the set of all probabilities is so large that many of us are not sure what it means to use a Dirichlet prior. For example, what about the tails of the prior? It was widely believed that a Dirichlet prior puts high probability on measures with tails close to the tails of the measure α. Indeed, the expected value of the probability of $(-\infty, t]$ is $\alpha(-\infty, t]/\alpha(-\infty, \infty)$. Recently, Hani Doss and Tom Selke, two Stanford students, determined that 20 years of intuition (and some published theorems!) were wrong. For example, let α be the standard Cauchy distribution. Then, with probability 1, the random measure picked from the Dirichlet with parameter α has moments of all orders. For further details, see Doss and Selke (1981).

The discussion above says nothing about where the fictitious data sequences come from. One suggestion is the use of typical sequences from a widely used probability model. This will be made specific in the next section.

2. Consistency of Bayes Rules

In ordinary finite dimensional problems, Bayes rules are frequentist consistent. This notion is of interest to Bayesians through the "what if" method.

Example 1. Dirichlet and Tail Free Priors

Let Θ be the set of all probability measures on the integers $0, 1, 2, \ldots, N$, where $N = \infty$ is allowed. For $p \in \Theta$, let p^∞ be the infinite product measure. Let π be a prior probability on Θ. Call the pair (p, π) consistent if for p^∞ almost all sequences $X_1, X_2, \ldots$ the posterior distribution of π given $X_1, \ldots, X_n$ converges to a point mass at p as $n \to \infty$. Measure theoretic details can be found in Freedman (1963). Consistency can be interpreted via the "what if" approach. For large n, a histogram of $X_1, \ldots, X_n$ will be close to the probability histogram of p, and one would hope the data would swamp the prior, forcing convergence to p.

The mental exercise becomes "what if X_1, X_2, ... is a typical sequence from p. Will the posterior get closer and closer to a point mass at p?"

Freedman (1963) proved that if N < ∞, then (p, π) is consistent if and only if p is in the support of π. This is intuitively the right result. In coin tossing, if the prior is supported in [$\frac{1}{2}$, 1] and the proportion of heads approaches $\frac{1}{4}$, the prior cannot converge to $\frac{1}{4}$. Freedman also showed that the finite dimensional result does not extend to N = ∞. Specifically, a prior π was constructed such that $p_{\frac{1}{2}}$, the geometric distribution with parameter $\frac{1}{2}$, was in the support of π, but for $p_{\frac{1}{2}}^\infty$ almost all sequences, the posterior converges to a point mass at a geometric distribution with parameter $\frac{3}{4}$. To surmount this problem, the notion of tail free prior was introduced. Briefly, a prior is put on Θ as follows. Pick a large k, specify the prior distribution of $\{p(i)\}_{i=1}^{k}$, leaving positive mass for $\{p(i)\}_{i=k+1}^{\infty}$, and complete the prior specification by "random stick breaking". This method includes the Dirichlet prior as a special case. It was shown that under very mild restrictions on the stick breaking part of the tail free prior π, (p, π) is consistent for all p ∈ Θ. Bayesian reaction to this phenomena was often something like "Oh! No one has a prior like that." The next example shows inconsistency in a natural setting.

Example 2. Robust Bayesian Estimation in the Symmetric
            Location Problem

Consider estimation of the length θ of an object after observing measurements $X_i = \theta + \varepsilon_i$, $1 \leq i \leq n$. The distribution of ε_i is symmetric about zero and of unspecified form. Bayesian treatment of this problem puts a prior on θ and ε;

$\theta \sim \mu(d\theta)$, $\varepsilon \sim \pi(d\varepsilon)$.

For examples and detailed discussion, see Chapters 3 and 4 in Box and Tiao (1973) or Dempster (1975). The prior on ε is often a partially specified standard family with 1 parameter for scale and 1 parameter for shape. A natural alternative is to use a Dirichlet or tail free prior, suitably symmetrized. See Dalal (1974).

The posterior for these priors is calculated in Diaconis and Freedman (1981) and in Doss (1982). When ε is given a Dirichlet prior with parameter measure θ, the mean of the marginal posterior distribution for θ is of the form

$$
\hat{\theta} = \frac{\int \theta w(\theta)\,d\theta + \sum_{i<j} \frac{x_i + x_j}{2} w_{ij}}{\int w(\theta)\,d\theta + \sum_{i<j} w_{ij}},
$$

with

$$
w(\theta) = \mu'(\theta)\Pi\alpha'(x_i - \theta),
$$

μ' and α' being densities of μ and α,

$$
w_{ij} = \alpha'\left(\frac{x_i - x_j}{2}\right)\mu'\left(\frac{x_i - x_j}{2}\right)\prod_{k\neq i,j} \alpha'\left(x_k - \frac{x_i + x_j}{2}\right).
$$

We were surprised to find that for some choices of α, the Bayes rule $\hat{\theta}$ need not be consistent. When α is Cauchy, we constructed bimodal densities, symmetric about θ, such that $\hat{\theta}$ oscillates between 2 wrong answers as the sample size gets larger and larger. This is closely related to the results showing that M estimates with redescending ψ functions are inconsistent in the symmetric location problem. See Freedman and Diaconis (1981). The result is surprising because, with the priors used, Bayes rules are consistent for estimating θ if ε is known, and for estimating ε when θ is known. There is a tendency to try to treat other parametric problems in a similar way. Put a Dirichlet prior on ε, and start computing. This example suggests caution. Doss (1982) has shown that very similar phenomena hold for both tail free and neutral to the right priors.

Here is another example of the "what if" method in this problem. Consider the Bayes rule $\hat{\theta}$ given above. What if one of the observations is moved very far from the others, will it have a profound effect on $\hat{\theta}$? The answer again depends on α. If α has exponential or longer tails, then $\hat{\theta}$ is robust in the sense of having good breakdown properties. See Diaconis and Freedman (1981) for details.

The results of the two "what if" questions suggest that α be log convex and have exponential tails. This leads to a Bayes estimator which, in large samples, had essentially the same influence curve as Huber's estimator. We next give some new results.

Example 3. Bayesian Numerical Analysis

Consider the following problem. Guess at a continuous function f on [0, 1], given the value of f at n points $\{f(x_i)\}_{i=1}^{n}$. The guess at f might be used to calculate $\int_0^1 f(x)\,dx$ or to interpolate f at an unobserved point. Several authors, Renyi and Palasti (1956), Wahba (1978) and Leonard (1981), have considered this problem in a Bayesian framework. The following results seem well-known parts of the folklore. If the prior is taken as Brownian motion, then the posterior mean

$$\hat{f}(t) = E\{f(t)\,|\,f(x_i)\}$$

is the straight line interpolant of the values $f(x_i)$. If the prior is taken as once integrated Brownian motion, then $\hat{f}$ is a cubic spline. If the prior is taken as k times integrated Brownian motion, then $\hat{f}$ is a spline of order 2k + 1. Of course, the whole posterior is available, not just its mean, and Grace Wahba has recently experimented with posterior confidence intervals for f. She has shown empirically that such intervals can have reasonable frequentist coverage properties. Perhaps this could be made rigorous by using techniques similar to those in Hinkley (1980) and Stein (1981).

These problems seem like natural candidates for the "what if" method. Suppose that $f(x_i) = 0$ for all i, where x_i runs through a dense grid. In the absence of very strong prior knowledge, one hopes that $\hat{f}(t)$ would converge to zero in some sense. We have shown that this is the case for k-times integrated Brownian motion, although a general theory, even for Gaussian processes, is lacking. The need for theory is suggested by the following construction, which produces a prior supported on a countable family of functions such that the zero function is in the support of the prior, and yet the posterior $\hat{f}$ does not converge to zero.

<u>Construction</u>. Define two sequences of bump functions g_n, h_n:

$g_n \sim \frac{1}{n}\Big\{$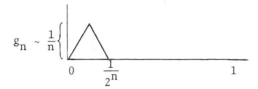

$h_n \sim 1\Big\{$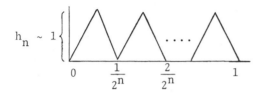

Thus g_n has a bump of height $\frac{1}{n}$ running from 0 to $\frac{1}{2^n}$, g_n being zero elsewhere, and h_n has periodic bumps of height 1. Put mass $\frac{c}{n^3}$ at g_n and mass $\frac{c}{n^2}$ at n_n, with c a normalizing constant. This defines a prior π on the continuous functions on $[0, 1]$. Observe that zero is in the support of π. Now consider observations over a grid at $\frac{1}{2^n}$. If all the observations are zero, then g_i and h_i, $1 \le i \le n-1$ are ruled out, so the posterior distribution is supported on g_i and h_i, $i \ge n$. Because of the choice of weights, the proportion of mass on h_i tends to 1. As a consequence, the Bayes estimate of a functional of the unknown f, such as

$$L(f) = \int_0^1 g(t)f(t)\,dt \quad \text{for some known g,}$$

converges to $L(\hat{f}) \to \frac{1}{2}\int_0^1 g(t)\,dt$, instead of zero.

The construction is given in only its barest form to reveal structure. It is easy to vary the prior to cause more colorful things to happen. For example, consider adding a Brownian motion to each g_i and h_i, forming

$$g_i(t) + \sigma_i B(t) \quad \text{and} \quad h_i(t) + \sigma_i B(t)$$

where σ_i tend rapidly to zero. The prior is taken as a countable mixture of these priors, with the same weights as

before.   Each piece of the mixture is consistent at any continuous function, yet the mixture goes awry at zero as before.   With a little more work, it seems possible to have a
prior which is inconsistent on a dense G_δ.

Example 4.   <u>Tail Free Priors Again</u>

Here is an example that shows that the $\frac{1}{2}$ - $\frac{1}{2}$ mixture of
a consistent prior and another prior can be inconsistent.   A
straightforward modification suggests that a countable mixture of Dirichlet priors can be inconsistent.

<u>Construction</u>.   Let Θ be all probabilities on the integers
$0, 1, 2, \dots$.   Let $\theta \in \Theta$ be a "log-Cauchy distribution"

$$\theta(i) = \frac{C}{i(\log i)^2}, \quad i \geq 10, \quad \theta(i) > 0 \quad \text{for} \quad \theta \leq i < 10.$$

Define θ^* by

$$\theta^*(i) > 0, \quad \theta^*(i) \neq \theta(i), \quad \theta < i < 10, \quad \theta^*(i) = \theta(i),$$

$$i \geq 10.$$

Take as a prior π, the prior generated by uniform stick
breaking.   Thus, the distribution of $p(0)$ is the same as a
point chosen at random on $[0, 1]$.   The distribution of $p(1)$
given $p(0)$ is the same as a point chosen at random on
$[0, 1 - p(0)]$, and so on.   Freedman (1963) shows that for any
$p \in \Theta$, (p, π) is consistent.   Consider now the prior

$$\pi^* = \frac{1}{2} \pi + \frac{1}{2} \delta_{\theta^*}, \text{ with } \delta_{\theta^*} \text{ a point mass at } \theta^*.$$

Since θ is in the support of π, θ is in the support of π^*.
Nevertheless, (θ, π) is consistent.   For θ^∞ almost all sequences $X_1, X_2, \dots$ the posterior distribution converges to a
point mass at θ^*.   The proof is somewhat delicate, and will
appear elsewhere.   One crucial feature of the construction:
both measures θ, θ^* have infinite entropy.   This forces a
super-exponential discrimination rate between a sequence
generated from θ and the measure π.   Because θ^* only differs
in finitely many places from θ, the discrimination rate is
exponential, and so dominates that of π.   For details, see
Diaconis and Freedman (1981).

3. Some Open Problems

The examples in Section 2 are not meant to suggest that it is hopeless to use Bayes rules in truly high-dimensional problems. Indeed, the work of Wahba and Leonard suggests that very useful, practical procedures can be developed this way. The examples do suggest that there is a good deal of work to do before statisticians have tried and true procedures to add to their tools. In this section we list some problems that we do not currently know how to solve.

(i) In the Bayesian approach to robustness discussed in Example 2, find a class of priors which is (a) supported on all symmetric distributions, (b) in some reasonable sense centered at the normal family, (c) robust, in the sense of having good breakdown properties, (d) consistent, (e) tractable, e.g., $\hat{\theta}$ is computable.

(ii) In the Bayesian approach to quadrature and interpolation discussed in Example 3, take standard procedures, such as Simpson's rule or quadratic splines, and determine if they are Bayes procedures for some prior on the continuous function on [0, 1].

(iii) In Examples 1-4, find sufficient conditions for a prior to be consistent at all parameter values in its support. The tail free priors are a class with this property, but a reasonable sufficient condition is lacking in Example 1, or more generally, for priors on all probabilities on the line. In Example 3, we do not know a nice condition for consistency of $\hat{f}(t)$ even for Gaussian process priors.

(iv) Measure-theoretic difficulties have been avoided up to now. They appear with real force if one tries to formulate a definition of consistency which includes both Examples 1 and 3. One way to approach a general definition of consistency is to consider a measurable space (X, B) and a family of probabilities $\{p_\theta\}_{\theta \in \Theta}$, with θ itself a measurable space. To model the sequential information coming in, it is natural to consider a decreasing sequence of σ-algebras $\mathcal{J}_n \subset B$. Let π be a prior on Θ. One possible definition of consistency is this: (θ, π) is consistent, if for each n there is a posterior distribution $\pi(\cdot | \mathcal{J}_n)$ and, except for a

set of p_θ probability 0, the measures $\pi(\cdot \mid \mathcal{J}_n)(x)$ converge to
a point mass at θ, weak star. If all the p_θ are equivalent,
then this definition captures what we want; indeed, under
reasonable restrictions, such as all spaces Polish, $p(\mid \mathcal{J}_n)$
is well defined and all versions of the posterior agree up
to a set of zero X measures.

In Example 3, the measures are all singular: here
$X = \Theta = C[0, 1]$ and $p_\theta(\cdot) = \delta_\theta$. In the first construction
of this example, measure theoretic difficulties are avoided
because the prior is atomic. In the second construction,
the prior was supported at a countable mixture of
$g_i(t) + \sigma_i B(t)$ and $h_i(t) + \sigma_i B(t)$. Now, there is a natural,
continuous version of the posterior, but it is also per-
fectly permissible to alter the posterior on a set of measure
zero, defining the posterior to put positive mass at the
function zero if $f(x_i) = 0$ for any x_i. The redefined posterior still works as a posterior, but now is consistent at
zero. We see this as more than a mere technicality.

REFERENCES

1. Box, G. E. P. and G. C. Tiao, Bayesian Inference in
    Statistical Analysis, Addison Wesley, Reading, MA, 1973.
2. Dalal, S. R., Nonparametric and robust Bayes estimation
    of location, Optimizing Methods in Statistics (J. S.
    Rustagi, ed.), Academic Press, New York, 1974, 141-166.
3. Dempster, A, P., A subjective look at robustness,
    Research Report S-33, Harvard University, 1975.
4. Diaconis P. and D. Freedman, Bayes rules for location
    problems, to appear in Proceedings of the Third Purdue
    Symposium on Statistical Decision Theory and Related
    Topics (S. Gupta, ed.), also Technical Report No. 177,
    Stanford University, 1981.
5. Diaconis, P. and D. Freedman, On inconsistent Bayes
    rules, unpublished manuscript, 1981.
6. Doss, H., Unpublished Ph.D. thesis, Stanford University,
    1981.
7. Doss, H. and T. Selke, The tails of probabilities chosen
    from a Dirichlet prior, Technical Report No. 178,
    Stanford University, 1981.

8. Efron, B., Comments on Blyth's paper, Ann. Math. Statist. <u>41</u> (1970), 1058.

9. Ferguson, T., Prior distributions on spaces of probability measures, Ann. Statist. <u>2</u> (1974), 615-629.

10. Freedman, D., On the asymptotic behavior of Bayes estimates in the discrete case, Ann. Math. Statist. <u>34</u> (1963), 1381-1403.

11. Freedman, D., On the asymptotic behavior in the discrete case II, Ann. Math. Statist. <u>35</u> (1965), 454-456.

12. Freedman, D. and P. Diaconis, On inconsistent M-estimates, to appear, Ann. Statist., also Technical Report No. 170, Stanford University, 1981.

13. Good, I. J., Probability and the Weighing of Evidence, Griffin, London, 1950.

14. Hinkley, D., Likelihood as approximate pivotal distribution, Biometrika <u>67</u> (1980), 287-292.

15. Leonard, T., Notes for a short course in Bayesian statistics, mimeographed notes, circulated at University of Wisconsin, 1981.

16. Renyi, A. and I. Palasti, On interpolation theory and the theory of games, MTA Mat. Kat. Int. Kozl <u>1</u> (1956), 529-540.

17. Stein, C., On the coverage probability of confidence sets based on a prior distribution, unpublished typed manuscript, 1981.

18. Wahba, G., Improper priors, spline smoothing and the problem of guarding against model errors in regression, J.R.S.S., Series B <u>40</u> (1978), 364-372.

P. Diaconis was partially supported by National Science Foundation Grant MCS79-08685.

Department of Statistics
Stanford University
Stanford, CA 94305

Department of Statistics
University of California at Berkeley
Berkeley, CA 94720

Purposes and Limitations
of Data Analysis

A. P. Dempster

1. INTRODUCTION

Data analysis is the central topic of the conference, as it is of the field of statistics. Data analysis is a broad subject, addressing varied questions, calling on different technologies, and facing important limitations. I propose to present brief analyses of several major dimensions of the subject, while raising questions and expressing opinions.

No one is likely to be pleased by an analysis of the foundations of a subject, least of all perhaps the author. My attempt results from a perception that the necessary elements of any realistic theory of data analysis need more explicit definition and more extensive dissection than appears to be available in print. Since the issues bear on appropriate choices for the development and practice of statistics, I hope that my analysis will encourage broader discussion of alternative futures for the field.

Some of the questions to be touched on below are: What are the relative merits of leading through exploratory data analysis vs. leading through models? When and what kinds of probability methods are needed in data analysis? How dependent is data analysis on formal inputs of prior knowledge for the purposes of answering the real world questions posed to the statistician by his client? What standards can be used to define a good data analysis?

Scientific Inference,
Data Analysis, and Robustness

2.   TECHNICAL VS. FUNCTIONAL STATISTICS

An important distinction which needs to be made at the outset contrasts technical statistics and functional statistics, where the former deals exclusively with tools and concepts which apply widely across many professions and sciences, while the latter requires blending adequate skill and depth in both technical statistics and substantive issues.

The discipline of statistics both theoretical and applied deals almost exclusively with technical statistics. The operating scenario can be roughly described as a sequence of processes: design for data collection, exploratory data analysis, modelling, inference, and decision analysis.   I define data analysis to consist of the middle sequence of steps between design and decision-making, wherein knowledge obtained from data is described and quantified.

A statistical analysis moves in order through these stages, but looks both backward and forward in the sense that the choices made at each step are informed both by the results of earlier stages and by an assessment of what will be needed at later stages.   Cycling back is generally necessary, and fortunately the need is widely recognized by careful data analysts.   More or less emphasis and more or less formality are awarded to each stage depending on the needs of the applied problem and the style of the analyst.

Formal analysis of the connection between technical statistics and the real world is almost nonexistent.   Discussion starts from specific cases and proceeds to largely uncodified guidelines for good practice, usually tied to fields of application.   Very little of this is taught to statisticians.

As a new Ph.D. twenty-five years ago, I was impressed by the neatness of the textbook picture of technical statistics but troubled by its apparently self-contained abstract structure, and I tried to ask about the "missing half" of statistics.   What was the justification for the assumed models which appeared then as now to carry so much of the burden in any application?   The obvious half derives from the immediate data, but these are insufficient, so the concealed half must come from prior knowledge of the substantive field, not knowledge in the sense of revealed truth, but rather knowledge of the current range of working methods, theories, and assimilated data.

It is also basic that the implicit background governs the form and detail of the substantive questions to be answered at the end of the technical processes.

To be functional, statisticians must approach problems from both technical and substantive sides.

3. RECOGNIZING STRUCTURE

One of the tensions in the field of data analysis is captured by the alternative terms <u>exploratory data analysis</u> and <u>statistical modelling</u>. I believe that the difference between EDA and SM is often exaggerated, since they share purposes and even share basic processes. Nevertheless, there is a major difference in that the SM approach often makes use of formal probability models whereas EDA typically excludes such models from consideration. One need not always pass from nonprobabilistic to probabilistic representations, and the step should be taken always with care. The step is often essential, however, because data analysis has goals beyond the limited purpose of reporting structure apparent in the data, and these goals usually require the formal representations of uncertainty conveyed by probability models.

The essence of EDA or SM is data reduction and manipulation so as to extract and exhibit comprehensible structure. Examples of structure are smooth approximations to empirical distributions or to empirical relations among variables. Such smooth forms are models. Structure, or models, may include sharp departures from smooth forms, such as deviant observations or peaks in an otherwise smooth spectrum.

Both EDA and SM cycle back and forth between model-fitting procedures and diagnostic model-checking procedures. EDA generally starts the process with summaries and displays, which means starting on the diagnostic side of the cycle rather than the fitting side. This makes sense as a rule because initial looks at the data may have a powerful veto

effect on large classes of models, thus reducing both the in-
efficiency of fitting and discarding obviously untenable
models, and reducing the danger of becoming model-bound. In
practice, however, if the cycling back and forth is pursued
with diligence and skill, the starting place should not have
a great effect on final reported results.

The term exploratory suggests creative attempts to un-
cover unsuspected new forms of structure which could be
crucial. Since creative efforts are only occasionally an un-
equivocal success, most expositions of EDA fall back on dis-
playing structure that is familiar to modellers who do not
stress exploratory aspects. Modellers may argue that the
cycling process provides ample opportunity to seek unsuspected structure.

Another theme suggested by the term exploratory is the
idea that EDA is guided by hypotheses about what will be seen
if specific reductions, manipulations and displays of the
data are carried out, whereas modellers are more concerned
with hypotheses about a larger world only partially described
by the data. I am impressed by the potential for interesting
discoveries in the use of computing power to produce long
sequences of two dimensional plots, such as many projections
of a high-dimensional data set, and of the potential for the
use of motion, depth illusions, color, and sound, to suggest
structure in data. Nevertheless, I believe that more humdrum
pictures showing relations which substantive theory suggests
are relevant are more likely to produce raw material for
productive models than are functionally unguided search procedures. Approaches from both ends are of course complementary, and much depends on the field of study, the quality of
the data, and both technical and functional skills of the
data analyst.

I now turn to the debate over the need or lack of need
for probability models. The data analysis school identified
with John Tukey appears to see little need for probability
models, while the modelling systems widely discussed by
George Box stress parametric probability models and associated probabilistic inference tools.

In part, the difference has to do with style more than substance. Both schools rely strongly on an implicit assumption that certain units, whether points in a data cloud or individuals in a sample, are a priori equally deserving of consideration in the analysis process. As a result the types of structure are often similar whether or not a formal probability sampling model is adopted. Evidence for the lack of difference is found in the ease with which proponents of exploratory analysis slip over to sampling models when they wish to practice confirmatory analysis. The line between empirical frequency displays and probability models is often fuzzy.

A virtue of EDA nonprobability modelling is the practice of concretizing models by identifying fitted models with representations of the data, preferably graphical representations, which stands in sharp contrast to the bad practice of adopting a probability model, fitting it by some reasonably efficient procedure, reporting a few canned tests of adequacy, and never really thinking hard or exploring for failures of the model which could compromise ultimate findings. There is something to the contention that the precision and elegance of many probability models may fool many users into acceptance without adequate safeguards. On the contrary side, there is the riposte that vital types of checking are not even possible without probability models. These are significance tests of goodness-of-fit and assessments of inference robustness, which I discuss further below.

Both probability models and nonprobability models can be used with enough skill and care to discover structure in data, and the natural advantages of one over the other vary with circumstance. I believe, for example, that there is no alternative to complex, many parameter models postulating several layers of probabilistic uncertainty for making sense of complex multilevel, multivariate, and time series data. Discussion of EDA vs. probability modelling as tools for penetrating complexity is a worthy topic, but beyond the scope of this paper.

A key reason for insisting on a role for formal sampling models is that they are required before one can pose and answer many well-defined real world questions, especially questions which motivated data collection and analysis in the first place. The questions are often difficult, and require integrating uncertain information from the data with uncertain information from other sources.

In summary, while the nonprobabilistic tools and skills of EDA are necessary for statistical practice, they are only sometimes sufficient. The statistician as reporter is an honorable profession, but many of us would rather see the statistician as problem solver. Many statistical problems should not be formulated and solved without probability models and associated confirmatory tools.

4. VARIETIES OF CONFIRMATORY ANALYSIS

One end point of data analysis may be numbers representing probabilistic assessments of uncertainty. The chief varieties are P-values, confidence coefficients, likelihood ratios, and posterior probabilities, including lower probabilities. Such outputs togehter with their interpretations constitute confirmatory analysis.

I use the term confirmatory with some hesitation, because current usage appears to lack precision on issues which I deem essential to a definition. A principal user of the term, namely John Tukey, appears to me to mean any valid use of significance tests or confidence statements, but his enthusiasm for EDA is not balanced by clear and forceful descriptions of CDA. By elaborating what the term conveys to me, I may encourage others to do likewise.

I wish to make three main points about confirmatory analysis. First, there is a clear separation of purposes between the use of CDA as a technical aid to EDA or modelling, and the use of CDA for the functional goals of answering substantive real world questions. Different purposes can call forth different varieties of techniques. Second, there is a fundamental distinction between confirmatory tools which interpret probabilities retrospectively, after the outcomes

to which they refer have been observed, and those which
interpret probabilities prospectively, to assess uncertainty
about outcomes which remain unknown. Both varieties of CDA
are firmly rooted in common sense reasoning about uncertainty.
Third, while the major body of statistical theory associated
with Neyman-Pearson-Wald provides important knowledge about
confirmatory techniques, especially techniques based on
retrospective interpretations of sampling probabilities, it
fails to treat the essential problem of data analysis, namely
what to think after the analysis is completed.

It is a standard part of EDA or SM to notice some
apparent structure in data, such as a linear trend or depar-
ture from linearity in a normal plot, and then to suspect
that the apparent structure may only be the result of a
chance fluctuation. Significance tests, despite the well-
known difficulties of interpretation when they are proposed
after seeing the data, as they usually must be in CDA, are
well-established and, I believe, legitimate modelling tools.
From Box (1980) one might gain the impression that such tests
are the sole component of the model-criticism phase of the
back and forth modelling process. It is therefore a useful
antidote to notice that Tukey's nonprobabilistic treatment of
modelling puts great stress on diagnostics. Many or perhaps
most widely used goodness-of-fit significance tests have
their origins in intuitively natural exploratory or descriptive statistics, and when inference-based principles such as
the likelihood-ratio principle are used to derive tests it is
generally comforting to obtain some geometrical or graphical
interpretation of the test statistic. In this fashion, CDA
provides back up for EDA or SM.

What holds true for testing also holds true for estimation. Serviceable techniques for many common model-fitting
situations are often derivable from nonprobabilistic EDA,
while inference-based techniques may pursue levels of efficiency not needed in practice. On the other hand, I believe
that likelihood-based techniques can often lead and promote
intuitions, especially in the modelling tasks created by an
increasingly data-rich world.

In contrast to the somewhat optional use of CDA to help
the statistician in judging what to report about apparent
structure in the data, there is a much less optional use in
functional problem-solving. For example, an experimenter who
has completed a randomized experiment may ask questions of
two kinds: Is there any evidence in the data that the treatment group differs from the control group? Or, how large is
a certain measure of the underlying treatment effect that
would have been produced in a much larger experiment? In
both kinds of question, uncertainty is of the essence.

There are of course technical parallels between the
modelling-reporting process and the external problem-solving
process, since the answers to the latter are derived from
models. For example, a fitted mean difference between two
samples may become an estimate of treatment effect. But the
underlying purposes of analysis are quite different, and the
functional responsibility taken on by the statistician as
problem-solver is so much greater that the choice and justification of confirmatory tools needs correspondingly greater
care.

The world of CDA is split down the middle between tools
which rely on retrospective (or postdictive) interpretations
of probability and prospective (or predictive) interpretations of probability. Parametric probability models nominally provide prospective measures of the uncertainty in prospective data to observers who know parameter values, and
margins of Bayesian models averaged over priors do the same
for observers who do not know the parameter values but do
subscribe to the priors. Once the data are in hand, these
prospective interpretations vanish and the same probabilities
can be interpreted only retrospectively, that is, as quantities whose smallness may produce surprise and thence skepti-
cism about the associated model. P-values, confidence coefficients, and likelihood-ratios are necessarily postdictive data analytic tools. They serve primarily to mark
off classes of models which are rendered dubious by the data,
from the rest which are not positively supported but only not
disconfirmed by the data.

The primary use of postdictively interpreted probabilities is in modelling, but by extension they are widely used to answer functional questions such as, "Is there an effect?". They cannot provide a negative answer, but sometimes provide a positive answer which scientists can accept.

Hard line Bayesian statisticians heap criticism on postdictive inference, which is indeed hard to defend against high standards of logical purity. But I have never perceived it necessary to recant from the position argued in Dempster (1971) that retrospective tools can be safely used for testing models, while prospective probability inferences are more crucial for assessing the values of unknowns. Box (1980) expresses a similar view. I would add that it may be possible to loosen the Bayesian straitjacket some by using belief functions or lower probabilities, as illustrated in Shafer (1982). And I would stress my belief that the emphasis on posterior prospective probability assessments is virtually a professional responsibility for a statistician who undertakes functional problem-solving.

In order to be clear about the large difference between CDA as described here and the image of statistical analysis conjured up by the widely taught frequentist theories of mathematical statistics, it is necessary to be explicit about what I mean by probability. A probability is a numerical measure of uncertainty about a specific outcome. That is, it is what is often called a subjective or personal probability, except that I propose we should all abandon the qualifiers as superfluous and misleading. A probability is a tool like an assumption, an estimate, or a decision. As with all tools, its place on a subjectivity/objectivity scale is determined by how much agreement exists among professionals concerning its appropriateness. A probability acquires objectivity from evidence either in the form of empirical facts or rational argument. Frequency is a concept closely related to probability, and serves both to calibrate probability and as a source of numerical probabilities. But long run frequency acquires the meaning of probability only sometimes, in the same way that measured distance is only sometimes a distance I will travel.

All this seems to me commonplace, and would not bear stating were it not for the fact that the most widely known theory of statistics makes a point of avoiding the concept of probability as just defined, and many proponents of the theory attack the scientific respectability of the concept, as though the determination of a probability differed in some fundamental way from the choice of a statistical technique or the specification of a model.

The Neyman-Pearson-Wald theory is closely related to confirmatory statistics. In fact, it is mainly a theory about techniques of confirmatory analysis. More precisely, it is a mathematically rich and an empirically illuminating theory about the long run performance of such techniques.

Neyman sought to elevate the theory about practice to a theory of practice by inventing the idea of inductive behavior whereby the statistician contemplates long run operating characteristics of possible decision rules, then selects a rule to be applied to the data. Taken literally, Neyman's theory suggests that all the statistician's judgment and intelligence is exercised before the data arrives, and afterwards it only remains to compute and report the results. Certainly, the theory does not address the statistician's thoughts processes once the data are in view. The theory is intellectually provocative, and an important chapter in the history of 20th century statistics, but since I view reasoning in the presence of data to be the crux of data analysis, I cannot take the theory seriously either as an actual or normative description of the practice of CDA. If we drop inductive behavior as a viable concept, then the Neyman-Pearson-Wald theory becomes an informal guide to CDA practice, less grandiose in scope but still very useful.

5. ROBUSTNESS AND PRIOR INFORMATION

It is close to twenty years since Box and Tiao (1962, 1964) introduced the study of inference robustness, or the sensitivity of Bayesian inferences to changes in assumed models, and contrasted inference robustness with criterion robustness, or the sensitivity of frequentist operating characteristics of procedures to changes in assumed models. Part of my thesis, not surprisingly, is that the former is essential to problem-solving functional data analysis, whereas the latter is beautiful theory but less relevant to practice.

Box (1980) contains a major statement on inference robustness, pointing out that the computing revolution makes extensive sensitivity analyses ever more feasible, and presenting a detailed example. Inference robustness has interesting parallels with diagnostic checking in that it is a basic type of data analysis which ideally should go through several iterations as different alternative models are scrutinized. Much greater effort to develop tools which will facilitate IRDA should have high priority.

Box (1980) very effectively stresses up front the mainspring of Bayesian robustness concerns: that models may fail without warning and drag down inferences with them. Accordingly, I once suggested modifying the concept of error of the second kind: "An error of the second kind would mean failure of a goodness-of-fit test to reject even though the ultimate analysis is critically sensitive to undetectable changes in the underlying model" (Dempster (1975)). Statisticians are all aware of the basic law of CDA which states that the size of a just detectable model failure varies inversely with the square root of sample size. Since the effects of sample size on the consequences of model failure are relatively minor, it follows that robustness is roughly independent of detectability. Yet the pressure and temptation are often strong to accept hypotheses when the lack of data protects one from being caught out. Clearly it is necessary to take further precautions if the credibility of posterior probabilities is to be protected.

The problem is: what to do? Box (1980) leaves unanswered questions on this score. His message suggests that if we follow his example we can robustify our analyses and be protected against danger. Perhaps unintentionally, he appears to suggest that we need only add a parameter or two to the model, and all will be well. But which few parameters should we add? If we add a few, why not a few more? And what about the questionable nature of conventional diffuse priors? Why not add many parameters with a tight prior?

What I believe is true and important, and is the real point that Box makes, is that one can achieve with judicious choices of additional parameters much if not all that the frequentist robustness school can usefully achieve, and do it better by matching up assumptions with corresponding estimators and producing side benefits like posterior judgments about outliers.

Locking up the frequentist dragon has not, however, freed the maiden. We merely come back to the essential problem of undetectable changes, or equivalently the problem of real limits on adaptive estimation when there is nothing perceptible in the data to adapt to.

A typical situation faced by a professional statistician is: first, a real world question is identified; second, data are collected which are obviously relevant to the question; third, from IRDA one finds that the answers to the question vary to a practically significant degree across plausible model variations which the data lack power to resolve. For concreteness, suppose we need to know the 1981 average medical expenditures of a specific class of residents of a town. A random sample of size 50 is drawn from a complete list, and an accurate determination is made for each member of the sample. In the course of IRDA studies it is learned that the posterior distribution of the population mean is disturbingly sensitive to a tail shape parameter which cannot be estimated with sufficient accuracy because there are too few sample points in the tail. Usually in practice the situations are more complex, but the type of questions is the same and the frequency of nonrobustness findings will surely go up as the practice of IRDA becomes more widespread.

What should be the response of the technical statistician? One response is to decline to serve the client on the grounds that adequate technology is unavailable. When the judgment of inadequate technology is sustainable, requiring

judgment on a case by case basis, then to decline is profes-
sionally responsible and should not be criticized. Another
response is to point out that a related question, such as
inference about a population median or a mean of expenditures
below a cutoff, may be answered much more robustly. Or to
point out that it is sometimes possible to robustify a design
using available tools, and to ameliorate subsequent analysis
problems thereby. These are valid responses and may be more
or less helpful depending on circumstance.

Another response is to risk advising a specific technique, possibly a Bayesian inference with a considered choice
of probabilistic model including a prior distribution, or a
non-Bayesian estimator justified by studies of frequentist
opearting characteristics. Both of these should come with
robustness documentation, the former from IRDA and the latter
from theoretical studies. I prefer and discuss only the
former because I see the statistician's task to be the recom-
mendation of posterior probability inferences and because I
believe that the formal introduction of prior information
from external sources is an important tool not to be discarded lightly, and a tool which has no counterpart on the
frequentist side other than informal and usually submerged
judgment.

An implication of my position is that a major priority
for technical statistics is research aimed at developing a
greatly expanded range of easily usable IRDA tools. I believe that many real world questions do have valid answers
which depend on statistical data analysis, but that outside
information sources are always involved and often critically
involved. We need data analysis tools with explicit hooks
for hanging on additional inputs of information. Similarly,
our tools should help to characterize the information in the
external sources, much as they do with immediate data sources.
We need much more emphasis on tools for combining different
data sources, both because of rapidly growing demands for
such tools, and because the potential for developing such
tools is great.

Finally, a few points on the narrower field of Bayesian robustness, stressing the connections between technical and functional issues. A sensitivity measure is a ratio of changes, like a derivative. The numerator is a measure of change in a posterior inference, while the denominator is a measure of change in a model. To interpret such a ratio it is necessary to establish functionally meaningful scales for both numerator and denominator. The numerator scale must establish something like a just seriously damaging change of inference while the denominator scale must specify some level of model change which is a plausible reflection of external sources of evidence. The necessity of functional inputs has to be stressed. Part of the impact of the examples in Box (1980) is lost because they are divorced from their real world settings, so that the numerator change is tied to model parameters instead of functional entities, while there is no way to think about prior inputs concerning undetectable serial correlation in a regression analysis or tail contamination in sampling without background knowledge tied to the specific example.

As hinted above, I am troubled by the idea that model-broadening should be grudging or parsimonious, to use terms appearing in Box (1980). The issue is one for the outside world, not for the statistician. Too often, perhaps, statisticians tackle data analysis where the real biases and model failures are much worse than conventional sensitivity analysis is likely to reach, but if they do tackle them, as I hope they will because of their unique command of important tools, they must be prepared to be much more technical about external information than is usual at present, and correspondingly allow much more complexity into parametric models. For example, in big regressions, or multivariate modelling generally, one should admit into the model large numbers of parameters, such as interactions in regression, much beyond the capacity of standard asymptotic inference tools to be helpful. Provided that reasonably tight priors on such parameters can be justified from the data or from external sources, then reasonably robust Bayesian inferences are possible despite the lack of parsimony.

6. QUALITY EVALUATION

How does one recognize a good statistical analysis?
Most evaluation of specific data analysis is anecdotal. An
example is briefly described, and the illustrated use of the
technique is pronounced successful. Critics find it all too
easy to shift the burden of proof by pronouncing that most
statistical analysis is useless or misleading. The situation
is paradoxical because data collection and analysis is a
basic tool for evaluation, so that statisticians should be
well equiped to construct systems for measuring and analyzing
their own performance.

One category of measures is process measures. Practice
should meet established professional and scientific standards.
Examples of the former are: quality of design and collection
procedures should be assessed according to standards which
are highly developed in many applied fields; computing should
strike an appropriate balance between accuracy and cost;
questions to be asked of the data should be explicitly formulated; formal tools should be introduced when needed, and
their absence should be explained and justified when not
needed; report material (written, tabular, and graphical)
should be effective and anticipatory of audience needs.
Scientific standards include: data should be analyzed in
several ways to check sensitivity of conclusions to methods;
unusual values or findings should be looked for, and studied
as to causes and to effects on the conclusions of the analysis; the appropriateness of methods should be judged in
relation to what can be seen in the data, what can plausibly
be assumed beyond the data, and what is only suspected or
even distantly possible; new questions and hypotheses should
be sought out and debated; alternative final conclusions
corresponding to alternative prior assessments should be
exhibited.

A second category of measures is product measures. The
best product measures are those which draw on verification
from follow-up data or replicated studies. One extreme is
illustrated by election-night forecasting where the follow-up

is quick and sharp. On the other hand, causal inferences from observational data may be difficult or impossible to verify. In such cases, evaluation by experts, despite its vulnerability to a mistaken consensus, may be the only immediate source of the end-point measures.

Statisticians have been much concerned, and rightly so, with the objectivity of their product. But, as Box pointed out in response to Barnard's (1980) call for "agreed probabilities", the statistical process requires a long sequence of judgments most of which are somewhat subjective. Consensus, whether it be on choices of techniques through inductive behavior, or on reported posterior probabilities, requires difficult discussions of standards and evaluations of case studies. Consensus on posterior probabilities is not meant to be easy, but progress must be achievable because, as Jack Good often remarks, inference is possible.

7. IMPLICATIONS

Two implications of my analysis stand out. The first is that much greater resources need to be brought to bear on computing problems which stand in the way of flexible applications of data analysis tools. Computing solutions require hardware and software, of course, but theoreticians are greatly needed for mathematical, numerical, and statistical analyses associated with algorithm development. Computing problems appear in nonprobabilistic EDA, in retrospective CDA, in prospective CDA, and in IRDA. The latter pair are as yet scarcely recognized as a field of research, but if my crystal ball is working the future depends heavily on feasible computing in these areas of data analysis.

The second implication is that the role of statistician as collaborator stressed by Box (1980) in his reply to discussants needs to penetrate more deeply into practice, and therefore into training. Instruction in statistics badly needs to convey not only technical content but a real sense of the functional contributions of data analysis technology to resolving borderline issues in many sciences and professions.

REFERENCES

1. Barnard, G. A., Discussion of Professor Box's paper, J. R. Statist. Soc. A 143 (1980), 404-406.
2. Box, G. E. P., Sampling and Bayes' Inference in scientific modelling and robustness, J. R. Statist. Soc. 143, Part 4 (1980), 383-404.
3. Box, G. E. P. and Tiao, G. C., A further look at robustness via Bayes's theorem, Biometrika 49 (1962), 419-432.
4. —————, A note on criterion robustness and inference robustness, Biometrika 51 (1964), 169-173.
5. Dempster, A. P., Model searching and estimation in the logic of inference, Foundations of Statistical Inference (ed. V. P. Godambe, D. A. Sprott), Hold, Rinehart and Winston, Toronto, 1971, 56-77.
6. —————, A subjectivist look at robustness, Bulletin of the International Statistical Institute 46, Book 1 (1975), 349-374.
7. Shafer, G., Lindley's paradox, J. Amer. Statist. Assoc. 77. (to appear - 1982).

The author was partially supported by national Science Foundation Grant MC77-27119.

Department of Statistics
Harvard University
Cambridge, MA   02138

Data Description

C. L. Mallows

ABSTRACT

Several recent developments give motivation for a study of description, divorced as far as possible from probability-based concepts. A good descriptive technique should be appropriate for its purpose; effective as a mode of communication, accurate, complete, and resistant. Some, but not all, of these attributes can be quantified. Design of powerful interactive statistical computing systems requires that attention be paid to providing effective output.

INTRODUCTION

A year ago Peter Walley and I [17] spoke at the annual statistical meetings in Houston, on "A Theory of Data Analysis?" with great emphasis on the punctuation, since we did not claim to be presenting such a theory, but rather to be outlining what problems such a theory might address, if and when it were to be developed, and to suggest a strategy of attack on these problems.

As one of the components of this proposed effort we suggested that there was a need and an opportunity for the development of a Theory of Description, and I'd now like to say something more about what I think such a theory might be concerned with. To start, I can do no better than to quote what Jimmie Savage said at the Conference on the future of statistics, held here in 1967 [1].

"I'm rather surprised to see myself acquiring a great respect for descriptive statistics. Everybody knows that descriptive statistics is strictly for Psych 100, and yet — though maybe you haven't used just that word for it — many of the most tantalizing things in statistical work today could be called

Scientific Inference,
Data Analysis, and Robustness

135

descriptive statistics. These are efforts to arrange and condense complicated bodies of data in ways that promise you a fighting chance to see what's essential. Factor analysis is one of the oldest theories of that sort. It has never been popular with statisticians, in part because the inference problems associated with it are so repulsive. But the factor analysts have something to stand on when they say, "The data are often abundant. This way of organizing it may give us an insight, and we ought not to wait for somebody to tell us the small probabilities that this insight is due to a fortuitous configuration of the data."

My thesis is that data description is important because it is a basic operation, more fundamental, it seems to me, than anything that can be captured in a formalism such as Bayesianism or the likelihood principle. Before thou canst sensibly postulate a probability model, thou hast to be able to examine and learn directly from data.

Several authors have discussed the use of formal significance levels as descriptive statistics. In his Presidential address to the Royal Statistical Society last March, David Cox said [2].

"Links between descriptive and probabilistically based statistics are important. In particular, there is need for some formal theory, even if very rough, in fields like cluster analysis which at present largely lack internal devices for assessing precision. Here methods based partly or primarily on computer simulation seem likely to be fruitful."

Nevertheless, it seems important to see how far we can go in developing a theory of description divorced as far as possible from the more familiar formal and probability-based statistical methods. Impetus for such a study of description comes from several sources.

First, at Bell Labs we seem to be frequently in the situation of wanting to deal with problems where no statistical model can be postulated in advance, and where pure description is completely satisfying. I shall give two examples shortly.

Second, John Tukey's advocacy of descriptive and exploratory methods. Tukey has presented many novel descriptive techniques, some of which are clearly useful, while others seem less compelling; but there is as yet no organized set of concepts with which we can discuss the merits of the procedures in detail.

Third, related to this, there has been for several years now a renewed interest in graphical methods of analysis and presentation.

Fourth, the proliferation of methods for the investigation of the structure of multivariate data - cluster analysis, multi-dimensional scaling and its extensions to three-way and higher dimensional structures, methods for skew-symmetric data matrices, etc. Most of these methods are as yet purely descriptive.

Fifth, a very recent interest in the development of systems for interactive statistical computing.

Sixth, developments in the theory of robust methods, some of which provide tools for the comparison of descriptive techniques without the necessity of referring to probability models.

Seventh, the need for methods of succinct presentation of the results of statistical analyses.

This last is perhaps the most important issue facing our profession. The responsibility lies more heavily on statisticians than on any other professional to make sure that their descriptions of the data not be subject to misinterpretation. This is what has given our discipline its bad name. We are all damned liars, because what we do is often presented obscurely and so does not convince. Of course some of what we do is obscure even to ourselves. A large study produces much output, especially in the exploratory stages. Much of this output is discarded, and what remains has varying degrees of precision and usefulness, which should be indicated in a published report.

I will be saying more about each of these topics as we go along. First however, I'd like to describe two studies in which description was found to be important.

The Corporations Study

The Corporations study was carried out at Bell Laboratories over a long period by Chen, Gnanadesikan and Kettenring. The first part of the work is described in [3]. The main purpose of the study was to determine a fair rate of return on investment for AT&T by finding other companies with "comparable risk" in line with a Supreme Court ruling that the rate of return allowed to a regulated company should be similar to that of companies with "comparable risk", which the Court did not define. The approach adopted in the study was to compile a list of variables thought to be related to risk, such as stock price variability, debt ratio, and other standard financial variables, and to look for companies most like AT&T in the values of these variables.

There also exists a classification of companies into Standard Industry Classes (electric utilities, oils, chemicals, etc.) which is presumably relevant to the problem. The analysis therefore fell somewhere between classification and clustering.

There were serious problems at an early stage of the study, in the selection and definition of appropriate variables. Subject-matter understanding, particularly of what the variables were actually measuring, and judgments of data relevance were needed to identify factors relevant to "risk" and to find suitable proxy variables to quantify these. A major challenge was to select a small number of economic and financial variables to adequately represent the many facets of company behavior. Technical problems of asymmetry, outliers and missing values also appeared at this stage. Further statistical problems in the analysis, concerning choice of appropriate metrics to reflect the shape of clusters and robust estimation of covariance matrices in the presence of multivariate outliers, gave rise to interesting theoretical developments.

Some conclusions of the study were that

(a)   AT&T falls in the middle of the industrial companies examined but in the tail of the utilities. (This is relevant because in some previous rate cases AT&T had been compared with electric utilities.)

(b)   The companies most similar (closest in the variable space) to AT&T generally had a much higher rate of return than AT&T.

Note that such conclusions are purely descriptive; they are based only on the observations that the companies appear to be grouped in a particular way and that features of the groupings appear to be relatively stable in time. It is hard to conceive of a plausible mechanism relating risk, which would have to be modelled as multidimensional, to the variables used, and equally hard to envisage a central role for classical statistical inference in this problem. Description is the best that can be done. The descriptive conclusions were moreover sufficient to achieve the purposes of the study.

The Bond Yields Study

The second study was undertaken by Denby, Kalotay and Kettenring [4]. The purpose was to understand variations in the market yields of AT&T and Operating Company bonds. When the study was undertaken, it was a controversial matter whether or not systematic differences among these yields exist. The data consisted of daily records of transactions in the N.Y. Stock Exchange; these data are extremely volatile and for many issues trading is infrequent.

After much analysis involving development of a suitable daily market index D_i and examination of many more complex models, they reached the simple model

$$y_{ijk} = D_i + C_j + Sx_{jk} + \epsilon_{ijk} \; ,$$

where y_{ijk} is the yield on day i of bond k of company j, C_j is the "company effect", and x_{jk} is the bond coupon.

This model describes the most important regularities in the data — in particular the company effect C_j, which quantifies risk (as perceived by investors) as associated with company j. This has uses in rate-case arguments and in deciding when a company ought to "call in" a particular bond.

The model's interpretation as a mechanism is more tentative. There is thought to be an effect associated with individual bonds, not all of which is accounted for by x_{jk}, which cannot be disentangled completely in this data from the company effect, especially for cases where there is only one bond for a company. The model is a first approximation to a mechanism.

Description and Exploration

I'd like to emphasize that it's not correct to identify "description" with "exploration". Cox points out in his address [2] that the splits between exploratory and confirmatory and between descriptive and probabilistic seem quite separate. Table 1 gives a classification of some techniques in each of three stages of an analysis, identification of patterns, fitting a model, and assessment of adequacy, with examples of descriptive and probability-based methods for each.

	Descriptive	Probabilistic
Identification of patterns	Scatter plots	Exploratory tests of independence
Fitting	Smoothing	Estimation, N.-P., likelihood, Bayes.
Assessment of adequacy	Diagnostic displays	Significance tests.

Table 1. Examples of Descriptive and
Probability-based techniques in each of three
stages of a statistical investigation.

Attributes of Descriptive Techniques

Let us turn now to the techniques of data description. In introductory texts there is usually a chapter on descriptive methods that describes histograms, a few measures of location and scale, and perhaps higher moments. But they give little basis for comparison, except relative to a probability model, once the concept of a sampling distribution has been developed. In his book John Tukey has presented a large number of novel descriptive methods. For example he suggests that a simple batch of numbers can usefully be described by giving a "five-number summary", consisting of the median, quartiles, and extremes. Associated with this is a suggested graphical technique called the Box Plot.

I believe this to be a good technique, but why is it good? What are the attributes of a good descriptive technique? Here is a short list of attributes that seem attractive, at least at first sight. A descriptive technique should be

Appropriate for the purposes of the study.

Effective as a mode of communication

Accurate

Complete

Resistant

Standardized.

Appropriateness is context dependent. This makes the concept very elusive — I will make only one comment. It seems to me that the fact that you are choosing to describe a batch of numbers as a whole, rather than to consider each of the elements of the batch separately, is connected with the basic nature of statistical reasoning. However, to me this choice does not seem to be directly connected with a probability concept, but is more primitive. For example, it seems meaningful to take (as Tukey does) the heights of fifty states and to summarize them as a single batch. It is also meaningful to look at the areas of 83 counties of Michigan. Yet it would not seem reasonable to attempt to summarize a batch of numbers some of which represent areas of counties while others represent heights of states. Note that in none of these cases is any probability concept involved. There is variability, but not randomness or uncertainty. Thus it seems to me that symmetry or exchangeability is fundamental but in a sense deeper than is captured by DeFinetti's development.

Let me move on to something a little less fuzzy.

Effectiveness. This attribute has several components, among which are familiarity, relative to the target audience, simplicity, and honesty.

Many of Tukey's techniques are simple indeed but lose effectiveness because they are not familiar. Excessive jargon impedes communication. My favorite example of this is of a visitor who gave a seminar at Bell Labs and referred to a cloud of points in three-space as being bread-shaped. This confused most of us until we remembered that the speaker was French so that for him bread was cigar-shaped.

The crucial issue is whether the attempted description communicates to its audience what was intended. This raises psychological issues. We are all familiar with what we can only assume are intentionally misleading graphical presentations by advertisers and politicians, but even when the more blatant distortions have been eliminated, communication can be poor, induced for example by the presence of "chartjunk" [6] which obscures the message, and psychological distortions.

There is a large literature on visual illusions, and of course pattern recognition has been studied intensively. But I believe that experiments that address questions of direct relevance to statistical graphics have been executed only very recently — my colleagues Cleveland, Harris, and McGill at Bell Labs and Diaconis at Stanford have begun work in this area [7-9].

Recently Tukey and Tukey have remarked [10] that we should not insist that our displays be "formally true" to the data they represent, rather that we should aim at true messages or accurate impressions from displays.

These attributes, appropriateness and effectiveness, compete with the remaining three, which are relatively more independent of context and audience — and so are more amenable to the development of a mathematical theory.

By *accuracy* I mean a measure of the closeness to which the description approximates the data. Formally, suppose we have a descriptor δ which takes a data — set x in a space X into a description d in a space D. To measure the accuracy of the description $d = \delta(x)$, we consider the inverse image $\delta^{-1}(d(x)) = A(x)$ which contains all the data sets y that have the same description as does the given data set x. Then d will be an accurate description of x if this set is small. So we postulate a measure of discrepancy $\rho(x,A)$ between x and a set of data sets A, which it is often convenient to take as being derived from a measure of distance $\rho(x,y)$ between two data sets x, y in X, thus:

$$\rho(x,A) = \sup_{y \in A} \rho(x,y) .$$

For example, we might measure the distance between two real-valued batches $\mathbf{x} = (x_1,...,x_n)$, $\mathbf{y} = (y_1,...,y_n)$ by the sum of the squares of the differences of the order-statistics:

$$\rho(\mathbf{x},\mathbf{y}) = \frac{1}{n} \Sigma \ (x_{i:n}-y_{i:n})^2 \qquad (*)$$

Then the accuracy of the description d of $\mathbf{x}$ can be defined as

$$\alpha_\delta(x) = \rho(x,\delta^{-1}(d(x))) \ .$$

Alternatively we could define the distance between a data set x and a set of data sets A by identifying a "most typical" member of A, namely that t such that $\sup_{z \in A} \rho(z,t)$ is minimized. It may or may not be appropriate to insist that $t \in A$. It turns out in many situations that such a t is not unique, so let $T(A)$ be the set of all t such that $\sup_{z \in A} \rho(z,t)$ is minimized. Then we can define the distance between x and A as being $\rho^*(x,A) = \rho(x,T(A))$, and the accuracy of the description of x is:

$$\alpha_\delta^*(x) = \rho(x,T(\delta^{-1}(d(x)))) \ .$$

Table 2 gives some calculations for the case of a batch of n real valued elements using the metric in $(*)$.

Table 2

Accuracies of some descriptions of a batch
of real-valued numbers $x_{(1)} \leq ... \leq x_{(n)}$

Descriptor	Description	Accuracy	
δ	d	$\alpha_\delta(d)$	$\alpha_\delta^*(d)$
median	Q_2	∞	∞
end-points	$(x_{(1)},x_{(n)})$	$\frac{n-2}{n}(x_{(n)}-x_{(1)})^2$	$\frac{n-2}{4n}(x_{(n)}-x_{(1)})^2$
3-no. summary	$(x_{(1)},Q_2,x_{(n)})$	$\frac{n-3}{2n}((x_{(n)}-Q_2)^2+(Q_2-x_{(1)})^2)$	$\frac{n-3}{8n}(()^2+()^2)$
5-no. summary	$(x_{(1)},Q_1,Q_2,Q_3,x_{(n)})$	$\frac{n-5}{4n}\overset{4}{\Sigma}(\text{gap})^2$	$\frac{n-5}{16n}\overset{4}{\Sigma}(\text{gap})^2$
mean, s.d.	$(\bar{x},s)$	$\frac{2(n-2)}{n}s^2$	$2\left[1-\frac{1}{\sqrt{n-1}}\right]s^2$

Completeness. This relates to the degree to which there is an absence of structure in the residual or undescribed variation in the data. "Accuracy" will increase with "completeness", but completeness not necessarily require great accuracy.

A beautiful example is given by Tukey ([5], p. 50). Figure 1 is the scatter plot for Lord Rayleigh's atomic weights of nitrogen, and the associated box plot. The scatter plot shows a clear clustering into two groups. Rayleigh noted that the measurements in one group came from nitrogen produced by decomposition of chemical compounds, while the other group came from samples of air from which the oxygen had been removed. The discrepancy led to the discovery of argon and eventually to a Nobel Prize. For this data set, the five number summary is fairly accurate, yet woefully incomplete.

FIGURE 1

(From J. W. Tukey, Exploratory Data Analysis
Addison-Wesley, Reading, Mass. Used by permission)

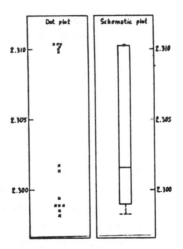

Rayleigh's 15 weights of "nitrogen"

A) DOT and SCHEMATIC PLOTS

For a less elementary example consider Ehrenberg's discussion [11] of "law-like" relationships such as that between height and weight of children.

The Heights and Weights Study

The data arose in a series of cross-sectional studies, each of which was concerned with the heights and weights of children in different age groups, for a particular locality, sex, epoch, race and social class.

Ehrenberg obtained the relation

$$\log w = .78h + .42 \pm .04 \tag{1}$$

which describes quite accurately all the data.

This may seem to be a simple regression problem amenable to standard statistical techniques. But Ehrenberg argues that classical statistical inference has no useful role here, since

(a) The form of (1) can be identified, and fitted adequately, without invoking statistical criteria.

(b) There are "real" deviations from linearity; that is, the within-group residual means are small (.01 or less) but, because of the size of the groups, significantly different from zero. Thus, the model (1) is *incomplete*. But these deviations cannot be related in any simple way to age, and do not seem to carry over to other localities, times, sex, etc.

(c) Our main interest in this relationship will concern its *generalizability*. This must be investigated by fitting the equation to height-weight data from different nationalities, sexes, and other plausibly relevant conditions. (It has been found to fit moderately well over a wide range of extraneous conditions.) Thus, the description of different but similar bodies of data may lead naturally to empirical generalizations, together with knowledge of their accuracy and range of validity.

In this study, unlike the Corporations example, we have a definite model for the data and can test its goodness of fit. However, since we are concerned with "law-like" behavior, the important problem is not to decide if departures from linearity might be "accidental", but rather to establish approximate *general* relationships — we must judge what regularities in the data are generalizable. The incomplete model is quite accurate, and is useful because it generalizes while a more complete model does not. In this case appropriateness is more important than completeness.

Resistance. This is the attribute of a descriptive technique that concerns the sensitivity of the description to small changes in the data — either small perturbations in all the data, or arbitrarily large perturbations in a small part of the data. Resistance is desirable, in that it often allows simple descriptions of the bulk of the data, which will usually need to be coupled with separate description of the rest of the data. To quantify it, we have concepts borrowed from robustness theory, namely the sensitivity function and the break-down point.

The sensitivity function is an empirical analogue of the influence function. It is simply the description function $\delta : X \to D$, considered as a function of one of its arguments, with the others held fixed. Several examples relating to measures of location appear in the Princeton study of robust estimates of location [12]. Devlin et al. [20] have explored the use of the concept as a descriptive tool., for example in the display of a scatter plot from which a correlation coefficient has been computed. Figure 2 is an example from [13] showing 50 points and contours of the sensitivity of the correlation coefficient. These contours help in describing the contribution of each data point to the computed correlation. Note that this procedure would not be effective if it were based on a resistant statistic. An effective description of such a

FIGURE 2

(From R. Gnanadesikan, Methods for Statistical
Data Analysis of Multivariate Observations
John Wiley & Sons, New York, N.Y. Used by
permission)

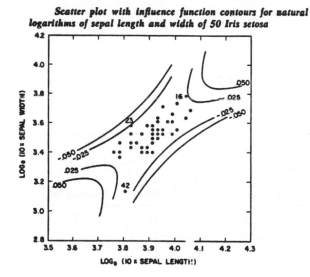

Scatter plot with influence function contours for natural logarithms of sepal length and width of 50 Iris setosa

scatter plot might consist of resistant measures of location, scale and correlation, together with a listing of outliers such as items Nos. 16, 23, and 42, which can be recognized through computation of sensitivity functions of non-resistant measures of location, scale and correlation.

Similarly we can define an empirical version of the breakdown point, which is itself a model-based concept. We define the EBP β as being the smallest fraction such that by changing no more than $n\beta$ data points we can change the description by an arbitrarily large amount. This definition implies that we have a metric on descriptions, but it would not be appropriate to use one derived from the accuracy we considered before, namely

$$\rho(d_1,d_2) = \sup_{x \in \delta^{-1}(d_1)} \sup_{y \in \delta^{-1}(d_2)} \rho(x,y) \ .$$

A suitable metric will be something like this. First, we need a notation to express the fraction of two data sets that agree, say

$$\text{agr}(\mathbf{x},\mathbf{y}) = \frac{1}{n} \# \{i : x_i = y_i\} \ .$$

Then we define the β-distance between two data-sets x,y as

$$\rho^{(\beta)}(x,y) = \inf_{\substack{x',y' \\ \text{agr}(x,x') \geq 1-\beta \\ \text{agr}(y,y') \geq 1-\beta}} \rho(x',y')$$

and the β-distance between x and a set A as

$$\rho^{(\beta)}(x,A) = \sup_{y \in A} \rho^{(\beta)}(x,y) \ .$$

Finally we define the β-accuracy of the description δ of x as being

$$\alpha_\delta^{(\beta)}(x) = \rho^{(\beta)}(x, \partial^{-1}(d(x))) \ .$$

With this definition in hand, we can characterize the description δ as being β-resistant if

$$\alpha_\delta^{(\beta)}(x) < \infty \ .$$

Some values are given in Table 3, using again the metric in (∗).

Descriptor	EBP $= \beta$	β-accuracy
(mean, s.d.)	0	$2s^2$
(γ-trimmed mean and s.d.)	γ	$2\dfrac{m(m-2)}{n(m-1)} s_\gamma^2 \quad (m = n(1-2\gamma))$
(median, quartiles)	1/4	$\dfrac{1}{4}((Q_2-Q_1)^2 + (Q_3-Q_2)^2)$

Table 3. Some Empirical Breakdown Points, and corresponding values of β-accuracy.

Similar definitions can be made involving "most typical" representatives of $\delta^{-1}(d)$, as we did before. Hodges [14] defined measures of "tolerance of extreme values" for a location estimator.

Standardization. Something I have omitted to mention so far is the issue of standardization of a descriptive technique. Consider an example discussed in Finch [15], which concerns the scores obtained in an examination by two groups of students who have been taught using different texts, say A and A'. Interest resides in the extent to which students who used one text did better than students who used the other text. The available data are the ranks of the students in the examination, which are

$$X = (3,5,7,8,11,14,15,17,18,22,24,25,26) \text{ text A}$$

$$X' = (1,2,4,6,9,10,12,13,16,19,20,21,23) \text{ text A}'$$

Consider several descriptions, which I give by listing ranks that they implicitly assign:

d_1 no difference, "random" $Y = Y' = (\alpha,2\alpha,...,13\alpha)$ where $\alpha = 27/14$

d_2 A better than A' $\qquad\quad Y = (14,15,...,26)$
$$\qquad\qquad\qquad\qquad\qquad\quad Y' = (1,2,...,13)$$

d_3 1 and 26 are special, for the rest there is no difference

$$Y \; = \; (1+\alpha,1+2\alpha,...,1+12\alpha,26)$$

$$Y' \; = \; (1,1+\alpha,1+2\alpha,...,1+12\alpha) \text{ where } \alpha = 25/13.$$

d_4 1,2,25,26 special, rest random

$$Y \; = \; (2+\alpha,2+2\alpha,...,2+11\alpha,25,26)$$

$$Y' \; = \; (1,2,2+\alpha,...,2+11\alpha) \text{ where } \alpha = 23/12.$$

Using the metric $\rho(x,d) = \dfrac{1}{n}\,\Sigma|x_i - y_i|$, we find the following numbers

$$d_1 \quad 1.5$$

$$d_2 \quad 5.0$$

$$d_3 \quad 0.69$$

$$d_4 \quad 0.58$$

These measures of accuracy have a direct meaning, and have no need of standardization.

Finch considers two ways of standardizing an accuracy measure such as $\rho(x,d)$. The first is called *descriptive power* and is the proportion of descriptions in a given class Δ that describe x more accurately than does d. The difficulty is in the choice of Δ, which to be meaningful must not include descriptions whose complexity is far greater than that of d. Typically, descriptors which might be considered as alternatives to d will differ greatly in complexity, interpretability and plausibility and will not therefore be comparable in this straightforward way.

A second method of standardization considered by Finch is to compute the "characterizing power" as the proportion of data-sets in some class Ω which are more accurately described by d than is x. A high value suggests that d is inadequate and that alternative descriptions should be sought, and a low value indicates that d captures some real feature of the data. This seems very similar to the usual interpretation of significance tests.

In the above numerical example, Finch takes Δ to be the set of all dichotomies, and Σ to be the set of all permutations of the scores. In this case he obtains descriptive power of our descriptor d_2 of about 0.999 and characterizing power about 0.86. I think he has made a slip, and that in this situation these two numbers should be equal, both being 0.86 in this case. I find these numbers less helpful than the direct measures of discrepancy presented above.

Computing

I'd like to end by saying something about developments in statistical computing. Most statistical software packages provide a number, ranging from small to moderately large, of options regarding the output that can be obtained and presented to the user. I believe there has been very little careful research into how useful these options are. Thus for example in a multiple linear regression package one can easily write down a dozen different plots that might be helpful, but who knows which are helpful in fact, and which are being used, and which should be the first to be incorporated into a statistical system?

The point is that we cannot allow a system to present all this information routinely. It would overwhelm the user. But equally, all these displays are potentially useful, and should be available when needed. Fortunately, there is an escape from this dilemma. We can use the computer itself to help us to find a good route for the analysis. This field is called 'expert software', which is a technical term in use in the artificial intelligence field. An example is a system for medical diagnosis called MYCIN. MYCIN produces diagnoses of infectious diseases, particularly blood infections and meningitis infections, and advises the physician on antibiotic therapies. The consultation is conducted in a stylized form of English; the physician provides the patient's history and laboratory test results. MYCIN has equaled the performance of human experts in its own limited field. An associated software system called TEIRESIAS develops the inferential rules that MYCIN applies, based on input from human experts.

Similar systems have been developed in a number of fields including geology, chemistry, and configuring computer systems.

In some of these systems a crude form of Bayes' theorem is applied to compute relative weights for the hypotheses currently under consideration.

The name 'expert software' is a little unfortunate in that it seems to suggest that once such a system is developed, human experts become superfluous. This is not at all the case. There is no suggestion that a software system could replace a human statistical consultant but rather that by incorporating and selectively applying a large body of factual information relating to available techniques, distributional results, and diagnostic technology, the system will facilitate effective and efficient analyses.

I'd like to quote from a paper by John Chambers presented at the recent Interface symposium [16].

> "The applications to medicine and data analysis have several important parallels and a few crucial differences. In both cases, the client is using the specialized knowledge of the expert software to provide analysis and recommendations to supplement the client's own knowledge. The software's role is advisory; it does make recommendations, but the client must make actual decisions.

> The differences might be summarized by saying that data analysis is a more heterogeneous, quantitative and iterative process than the form of medical diagnosis modeled by MYCIN. The diagnostic results in MYCIN are provided by the client; in our case, they will include both information from the client and, more frequently, the results of extensive statistical calculations."

In addition to the difficulty of designing the software to implement such a system, which is an area that is currently under intensive development, we have several purely statistical problems. Many of the existing expert systems are sufficiently modular that they can be switched to handle a new subject matter area simply by reading in a new set of action rules. Achieving a similar degree of modularity in the statistical context would be very difficult, the most we can hope for is that the statistical system be modular with respect to fields of application. More to the point of my topic today, the design of appropriate descriptions of intermediate results presents an enormous challenge. I think it entirely reasonable, for example, to expect that an expert system would draw the users' attention to a peculiarity such as that of Rayleigh's weights. The system should also be able to present for consideration several alternative descriptions in parallel.

Some of my colleagues are currently trying to design an expert system in the limited domain of multiple linear regression.

Conclusion

George Box has been telling us [18] that the time has come to put to rest the old controversies between Bayesianism and frequenticism — the two approaches deal with different problems, and are complementary. It seems to me that there are other issues that complement these in two further directions. First, the challenges that arise from dealing with really large data-sets, where delicate questions of significance or efficiency are unimportant. The problem is that of finding meaningful structure in the first place. And second, the opportunities that arise from using the power of the computer and its associated graphical peripherals.

Nothing in the formal theories that we are familiar with bear on these problems, and I think it is delightful to realize that there are enough challenges to keep us all busy.

REFERENCES

[1]   Savage, L. J. (1968). Discussion on Statistical Inference. In *The Future of Statistics*, ed. D. G. Watts. Academic Press, New York, (pp. 146-147).

[2]   Cox, D. R. (1981). Theory and general principle in statistics, J. Roy. Statist. Soc. A **144**, 289-297.

[3]   Chen, H. J., Gnanadesikan, R. and Kettenring, J. R. (1974). Statistical methods for grouping corporations, Sankhya B **36**, 1-28.

[4]   Denby, L., Kalotay, A. J. and Kettenring, J. R., (1978). A statistical study of market yields of Bell System bonds. Proceedings of the Business and Economics Section, American Statistical Association, 395-400.

[5]   Tukey, J. W. (1977). *Exploratory Data Analysis*. Addison-Wesley, Reading, MA.

[6]   Tufte, E. (1978). Data graphics. Paper presented at the First General Conference on Social Graphics, Leesburg, Virginia.

[7] Cleveland, W. S., Harris, C. S., and McGill, R. (1981). Circle sizes for thematic maps. Submitted for publication.

[8] Cleveland, W. S. and McGill, R. (1981). Color induced distortion in graphical displays. Submitted for publication.

[9] Cleveland, W. S. and Diaconis, P. and McGill R. (1981). Variables on scatterplots look more correlated when the scales are increased. Submitted for publication.

[10] Tukey, J. W. and Tukey, P. A. (1981). Graphical display of data sets in 3 or more dimensions. In *Interpreting Multivariate Data*, ed. V. Barnett. Wiley, London.

[11] Ehrenberg, A. S. C. (1968). The Elements of Law-like Relationships. J. Roy. Statist. Soc. A **131**, 280-302.

[12] Andrews, D. F. et al. (1972). *Robust Estimates of Location*. Princeton University Press.

[13] Gnanadesikan, R. (1977). *Methods for Statistical Data Analysis of Multivariate Observations*. Wiley, New York.

[14] Hodges, J. L. Jr (1967). Efficiency in normal samples and tolerance of extreme values for some estimates of location. Proc. 5-th Berkeley Symp., Vol. I, 163-186.

[15] Finch, P. D. (1979). Description and analogy in the practice of statistics. Biometrika **66**, 195-208.

[16] Chambers, J. M. (1981). Some thoughts on expert software. Presented at Computer Science and Statistics: 13th Symposium at the Interface.

[17] Mallows, C. L. and Walley, P. (1980). A theory of data-analysis? Proceedings of the Business and Economics Statistics Section, American Statistical Association, 8-14.

[18] DeFinetti, B. (1974). *Theory of Probability*, Vol. I. Wiley, New York.

[19] Box, G. E. P. (1980). Sampling and Bayes' inference in scientific modelling and robustness. J. Roy Statist. Soc. A, **143**, 383-430.

[20] Devlin, S. J., Granadesikan, R., and Kettenring, J. R. (1975). Robust estimation and outlier detection with correlation coefficients. Biometrika **62**, 531-545.

Bell Laboratories
Murray Hill, New Jersey 07974

Likelihood, Shape,
and Adaptive Inference

David F. Andrews

1. INTRODUCTION

The analysis of scientific observations is a complex,
sequential procedure involving many steps and many decisions.
The selection of observations to study, the selection of a
'dependent' variable, the initial transformations of the data,
the initial summarization of the data, all of these choices
are based on the understanding of the present experiment and
on experience with related data. Typically, no formal model
is used in making these selections although each of these
decisions could be formulated as a statistical problem. Thus
a substantial component of any statistical analysis is sub-
jectively decided by the analyst.

It is often preferable to make a simple model of summaries
of the data rather than to create a comprehensive model describing all of the available data. Simple models are typically
  i. easier to understand, and
 ii. easier to implement.
For a further discussion see for example Cox and Hinkley (1974
pages 3, 4). Often small precise models are preferrred to
larger, more correct models.

Distributional form is a common component of statistical
models. In cases where this form is not known exactly (the
usual cases) there are two basic approaches.

Scientific Inference,
Data Analysis, and Robustness

i.   Likelihood base methods: add sufficient parameters to
     the model to allow for differences in form, or

ii. Robust methods: use procedures which work well under a
     variety of shapes including those parameterized in i).

The question is whether distributional form should be formally
or informally included in the statistical analysis. Some
methods (Box (1980), Fraser (1976)) follow i), while others
(Mosteller and Tukey (1977)) follow ii). Each set of methods
has its advantages and disadvantages. The choice is based on
the relative properties and characteristics of each method.

The quantitative differences among the estimates derived
using different methods are likely to be small. For example,
Box (1980) suggests that a posterior mean has a form similar
to an M-estimate. The theoretical frameworks from which methods
may be derived may differ, but framework is not an important
characteristic of a method. Rather methods are characteristics
of theories. The important differences include

 i. feasibility and

ii. information about shape.

Concerning feasibility, software exists for the routine use of
M-estimates in general linear models. The same is not true
for likelihood based methods. For even the simplest problems,
likelihood calculations are awkward.

Concerning the information about shape, it is hoped that
the inclusion of shape parameters will lead to useful information about this shape. It is of interest to assess the
amount of information about shape in typical problems.

Much has been written about adaptive estimators. The
motivation for these in part is that, from samples of sufficient size, information about shape should be available in
order to explicitly influence the form of the estimator. It
is doubtful that much useful information about shape is available from samples of size 1 or 2 or 4 for example. How large
do samples have to be to provide useful information about
shape? Arguments based on relative error suggest that samples
of size 100 are required to estimate a second moment with
reasonable precision. Samples of size 1000 are required to

reliably estimate a third moment, the first moment to contain information about shape. These arguments suggest then that there is only limited information about shape available from small samples.

On the other hand, Relles and Rogers (1977) show that for samples of size 20 a Bayesian estimator has high efficiency relative to a related estimate based on a fixed shape.

This paper reports on a study of a particular problem, the k-sample problem. Some characteristics of two likelihood methods are compared. Section 2 presents some basic definitions and notation and applies these to the k-sample problem. Section 3 summarizes the results.

2. DEFINITIONS

Consider n independent observations:

$$y_i \quad i=0,1,\ldots,n-1$$

with structure of k samples of the same size (n, a multiple of k):

$$y_i = \mu_{\left\lceil \frac{ki}{n} \right\rceil} + \sigma e_i$$

where μ is a vector of location parameters, σ is a scale parameter, and where e has a t distribution with λ degrees of freedom. The case $k = 1$ corresponds to the estimation of a simple mean.

Let $f_\lambda(e)$ denote the density of e. Following Box and Cox (1964), we define the maximized likelihood

$$\text{Lmax}(\lambda) = \max_{\mu} {}_\sigma \Sigma \ln \sigma^{-1} f_\lambda((y_i - \mu_{\left\lceil \frac{ki}{n} \right\rceil}) / \sigma).$$

Following Fraser (1976), we define the marginal likelihood

$$\text{Lmarg}(\lambda) = \ln \int_\sigma \int_\mu \Pi \sigma^{-1} f_\lambda((y_i - \mu_{\left\lceil \frac{ki}{n} \right\rceil}) / \sigma) d\mu \, d\sigma.$$

The maximization of the definition of Lmax was calculated using the Nelder-Mead procedure of Hill (1973). The iterated integrals in the definition of Lmarg were evaluated using Gaussian quadrature after a preliminary transformation of variables.

Both likelihoods were evaluated for 11 values of λ, $\log(\lambda)$
equally spaced on the interval $[\log(1), \log(40)]$. Data were
simulated using degrees of freedom $\lambda_0 = 2$ and 6. The number
of samples, k, were fixed at 1, 2 and 4. Results for k = 1
and 4 are described in detail here.

The shape of both likelihoods were summarized by

$$d_1 = L(\lambda_0) - L(1) \quad \text{and} \quad d_{40} = L(\lambda_0) - L(40)$$

where λ_0 represents the degrees of freedom used to generate
the sample.

3. RESULTS

Figure 1 presents the first 6 maximized likelihood functions Lmax for $\lambda_0 = 2$. In this display the three curves
represent likelihoods derived from 1, 2, and 4 sample problems.
Since the likelihood is maximized over an increasingly complex
model, the value of the maximized likelihoods lie above each
other, with the likelihood corresponding to the 4-sample
problem on top. Note that there is no consistent pattern of
these functions having maxima at λ_0. The three curves
corresponding to k = 1,2,4 show surprising differences even
though these functions are based on the same deviates with
only different models.

To study the shape of the likelihoods under different
conditions, 100 likelihoods were calculated from the same set
of uniform deviates. The data generated here have the structure of a 2^4 factorial experiment:

$$\begin{bmatrix} \text{Lmax} \\ \text{Lmarg} \end{bmatrix} \times \begin{bmatrix} \lambda_0 = 2 \\ \lambda_0 = 6 \end{bmatrix} \times \begin{bmatrix} k = 1 \\ k = 4 \end{bmatrix} \times \begin{bmatrix} d_1 \\ d_{40} \end{bmatrix}$$

The likelihood functions were derived from 100 samples of
size 20 transformed to yield t_2 and t_8 variates. Figure 2
consists of plots of d_1 vs d_{40} for the 8 possible combinations of the 3 other factors.

Points in the region $d_1 > 2$, $d_{40} > 2$ correspond to
likelihoods clearly indicating the true shape λ_0. Notice the
number of such likelihood functions. Notice the number of
functions for which $d_1 < 0$ and $d_{40} < 0$. These correspond

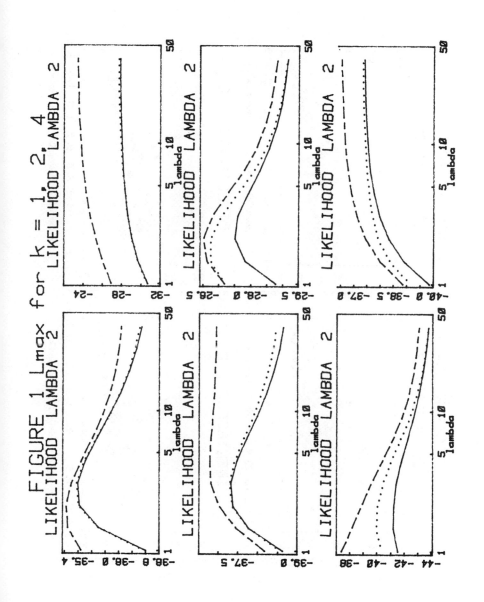

FIGURE 1 Lmax for k = 1, 2, 4

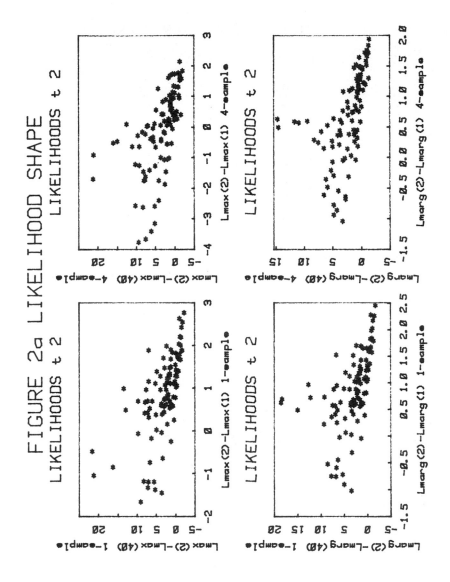

FIGURE 2a LIKELIHOOD SHAPE

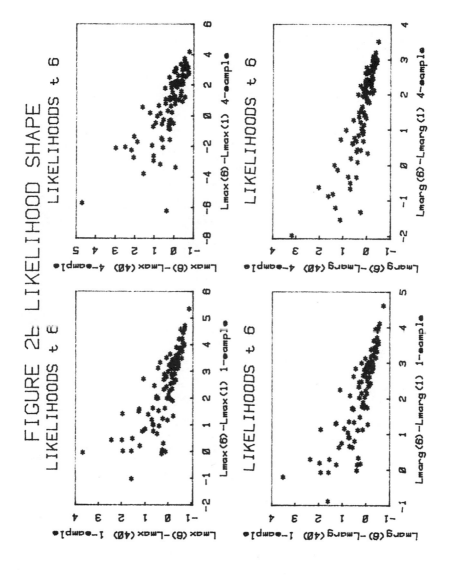

FIGURE 2b LIKELIHOOD SHAPE

to functions for which 1 or 40 degrees of freedom yield values
of the likelihood function higher than the value at λ_0, indicating that an extreme value of λ is more plausible than the
true value λ_0.

Figure 3 consists of plots of Lmax vs. Lmarg. Notice
the surprising agreement under some conditions. Notice also
the surprising disagreement for k = 4.

4. DISCUSSION

The calculations presented here are delicate. Both forms
of the likelihoods required several days of program development. The procedures developed are peculiar to the k - sample
model. Different problems would require more work. (The k -
sample problem was one of few for which marginal likelihoods
may be calculated by numerical integration.) In general, the
marginal likelihood requires a k + 1 dimensional integral.
Quadrature methods are not useful for k > 3 parameters.
Efficient Monte Carlo methods may be developed.

The differences between marginal and maximized likelihood
methods appear to increase with k (for fixed n). The maximized likelihood appears to be less informative. Unfortunately
the cases of large k are precisely the cases where quadrature
calculation of Lmarg is awkward.

The maximized likelihood Lmax was found to be sensitive
to rounding of the data.

For this sample size, the information available about λ
is not precise. Few samples give likelihoods which rule out
both extreme values of λ, (1 and 40). Indeed many samples
indicate that an extreme value of λ is more likely. The
estimators (and related inferences) associated with these
likelihoods may have good properties, -- but so would other
estimators.

The important differences between likelihood and robust
methods involve feasibility of calculation and information
about shape. In this problem, calculation of robust estimates
may be simply done using iterated weighted least-squares while
likelihood evaluation is awkward. There is no reliable information about shape available in this problem. Robust estimates may be calculated in more general linear models.

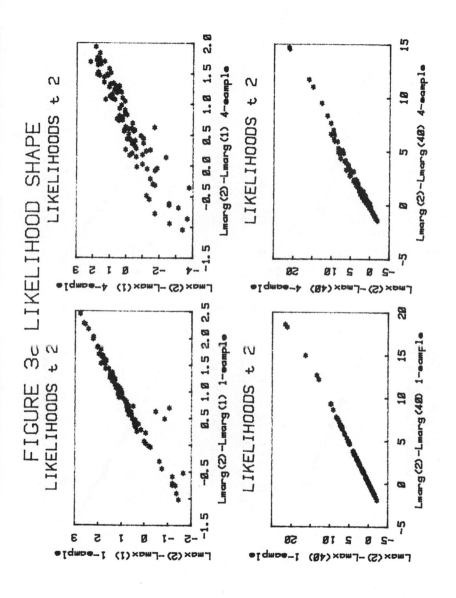

FIGURE 3c LIKELIHOOD SHAPE

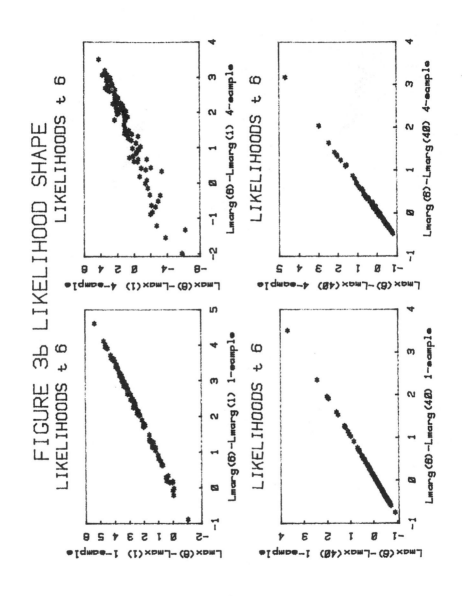

FIGURE 3b LIKELIHOOD SHAPE

REFERENCES

1. Box, G.E.P. (1980). Sampling and Bayes' inference in
    scientific modelling and robustness (with Discussion).
    J. R. Statist. Soc. A, <u>143</u>, 383-430.

2. Cox, D.R., and Hinkley, D.V. (1974). <u>Theoretical Sta-
    tistics</u>. Chapman Hall, London.

3. Fraser, D.A.S. (1976). Necessary analysis and adaptive
    inference. J. Amer. Statist. Ass., <u>71</u>, 99-113.

4. Mosteller, F. and Tukey, J.W. (1977). <u>Data Analysis and
    Regression</u>. Addison-Wesley, Reading, Mass.

5. Relles, D.A. and Rogers, W.H. (1977). Statisticians are
    fairly robust estimators of location. <u>J. Amer. Statist.
    Ass</u>. <u>72</u>, 107-111.

This research was supported in part by the National Science
and Engineering Research Council of Canada.

Department of Statistics
University of Toronto
Toronto, Ontario
Canada M5S 1A1

Statistical Inference
and Measurement of Entropy

Hirotugu Akaike

1. INTRODUCTION

It is a historical mistake that statisticians consistently ignored the fundamental contribution to the discipline made by a physicist Ludwig Boltzmann. Boltzmann [10] developed an interpretation of the thermodynamic entropy as the logarithm of the probability of a molecular distribution which defines the thermodynamic state of the system. This interpretation is best known through the later formulation by Planck [21, p.556]

$$S = k \log W + const,$$

where S denotes the entropy, k the Boltzmann constant and W the probability of the state. Boltzmann's demonstration of the thermodynamic irreversibility, the second law of thermodynamics, is based on this interpretation and is considered to be one of the great achievements of nineteenth century science [17].

The significance of the scientific implication of this formula can easily be confirmed by the fact that Planck [21] literally based his quantum theory of radiantion on this formula and Einstein [12], who called the content of the formula by the name of Boltzmann's principle, developed his quantum theory of light by freely applying the principle.

Great ideas in statistics have seldom been unrelated with the concept of entropy. With a proper definition of entropy, to be given in the next section, the chi-square goodness of fit test of Pearson [20] is seen to be directly concerned with the measurement of an entropy. The absolute criterion of fit of Fisher [13], which is the log likelihood of the distribution in the present terminology, can be viewed as a natural estimate of an entropy. We can also see that log likelihood ratio test statistics are concerned with the measurement of entropies. Even a justification of the use of Bayesian models can easily be obtained with the aid of an entropy.

Certainly there are enormous number of methodological papers on statistics which deal with some sort of statistical concept of entropy. Nevertheless, it is hard to find explicit references to Boltzmann's original work. This looks rather curious when we notice that Ramsey [22, p.64] simply took it for granted that there is a kind of probability whose logarithm is the entropy. By neglecting the Boltzmann's contribution statistics abandoned the privilege of utilizing the historically most successful interpreation of a basic statistical quantity.

Based on the recognition that the probabilistic interpretation of entropy is the key concept that connects statistics with probability the present author has been advocating the entropy maximization principle [2]. This is the principle that describes the purpose of statistics as the production of procedures to maximize the expected entropy. The purpose of the present paper is to show by using a collection of elementary examples how statistics can gain scientific consistency and practical productivity by adopting this point of view. The basic problem here is the measurement of entropies.

2.   ENTROPY AND LIKELIHOOD

The probability of getting a frequency distribution $(n_1, n_2, \ldots, n_k)$ by a repeated sampling from a population specified by the probability distribution $p = (p_1, p_2, \ldots, p_k)$ is given by

$$W = \frac{n!}{n_1! \, n_2! \, \ldots \, n_k!} \; p_1^{n_1} \, p_2^{n_2} \, \ldots \, p_k^{n_k} \; ,$$

where $n = n_1 + n_2 + \ldots + n_k$. By using the relation $\log n! = n \log n - n + o(n)$ we get $\log W = n\, B(r;p) + o(n)$, where $B(r;p)$ is defined by

$$B(r;p) = - \sum_{i=1}^{k} r_i \log \left(\frac{r_i}{p_i}\right) \; ,$$

where r stands for the distribution $(r_1, r_2, \ldots, r_k)$ defined with $r_i = n_i/n$. If in fact the frequency distribution is obtained by sampling from a population specified by the probability distribution $q = (q_1, q_2, \ldots, q_k)$, $B(r;p)$ will converge, when n is increased indefinitely, to the entropy of the true distribution q with respect to p defined by

$$B(q;p) = - \sum_{i=1}^{k} q_i \log \left(\frac{q_i}{p_i}\right).$$

The present definition of entropy retains Boltzmann's original interpretation of entropy as the logarithm of the "probability" of getting the true distribution q by sampling from the assumed distribution p and thus will be a natural measure of the deviation of the distribution q from p. Since $0 \le W \le 1$ we have $-\infty \le B(q;p) \le 0$ and $B(q;p) = 0$ only when $p = q$. When the probability distribution q and p are respectively defined by the density functions $q(x)$ and $p(x)$ with respect to a measure dx, the entropy is defined by

$$B(q;p) = - \int q(x) \log \left(\frac{q(x)}{p(x)}\right) dx.$$

The quantity defined by $I(q;p) = -B(q;p)$ is called discrimination information by Kullback [18]. Good [15] suggests to call it expected weight of evidence. Historically it would be quite appropriate to call it simply negentropy.

From the definition of entropy we have

$$B(q;p) = E_x \log p(x) - E_x \log q(x),$$

where E_x denotes the expectation with respect to the true distribution $q(x)$. When there are several models specified by the distributions $p_m(x)$ ($m = 1, 2, \ldots, M$) the best approximation to $q(x)$ is defined by the one with maximum $B(q;p_m)$. Since $E_x \log q(x)$ is common to all the $B(q;p_m)$'s the best approximation can be obtained by searching for the maximum of $E_x \log p_m(x)$. When an observation z of x is obtained $\log p_m(z)$ provides a natural estimate of $E_x \log p_m(x)$.

Thus we can see that for the purpose of comparison of the models we can use the log likelihood $\log p_m(z)$ as an estimate of the entropy $B(q;p_m)$ <u>without any specification of the "true" q.</u>

The present observation provides an objective justification of the use of likelihood as a "measure of rational belief" [14 p.68]. The method of maximum likelihood may thus be considered as a procedure of maximizing the entropy through the measurement of entropies by log likelihoods.

3. TEST AS MEASUREMENT OF ENTROPY

Boltzmann's work on entropy is essentially related with a characterization of the energy distribution as a distribution of exponential type. Thus the use of the probabilistic definition of entropy in statistics can best be explained with exponential family of distributions. Here we consider the exponential family defined by

$$p(x|\theta) = u(x) \exp\{s_0(\theta)t_0(x) + s_1(\theta)t_1(x) + \cdots$$
$$+ s_h(\theta)t_h(x)\}$$

with $t_0(x) = 1$. When an observation z is obtained we have

$$B\{p(\cdot|\theta_*); p(\cdot|\theta)\} = \sum_{i=0}^{h} (s_i(\theta) - s_i(\theta_*))\, t_i(z),$$

where θ_* denotes the maximum likelihood estimate, and we get the relation [19,23]

$$\frac{p(z|\theta)}{p(z|\theta_*)} = \exp [B\{p(\cdot|\theta_*); p(\cdot|\theta)\}].$$

This result shows that the likelihood $p(z|\theta)$ is proportional to the "probability" of getting the "true" distribution $p(\cdot|\theta_*)$ by sampling from the distribution $p(\cdot|\theta)$, where "probability" means that it is defined as the exponential of an entropy. Thus we see that with the use of the family the usually implicit true distribution automatically obtains its explicit representation through $p(\cdot|\theta_*)$. A somewhat trivial but important conclusion from this observation is that only the characteristics of data z that can be refected in $p(\cdot|\theta_*)$ can be handled by the use of this family. This means that the definition of a true distribution is always relative to the assumed model.

The thermodynamic entropy is a relative concept in the sense that only the difference has a physical meaning. This relative property is inherited by the log likelihood. The present result shows that the relativistic nature is an essential characteristic of the statistical inference realized through the use of a model or family of distributions.

For the case of a multinomial family $p(x|\theta)$, where $x = (x_1, x_2, \ldots, x_k)$ and x_i denotes the frequency of the i-th category within the sample of size n from a population defined by $\theta = (\theta_1, \theta_2, \ldots, \theta_k)$ $(0 < \theta_i < 1$ and $\Sigma\, \theta_i = 1)$, for an observation z of x we get $\theta_* = r = (r_1, r_2, \ldots, r_k)$ with $r_i = z_i/n$ and we have

$$B\{p(\cdot|\theta_*);p(\cdot|\theta)\} = n\ B(r;\theta).$$

When the data is actually sampled from a distribution defined with θ we have

$$\log\left(\frac{r_i}{\theta_i}\right) = \frac{r_i - \theta_i}{\theta_i} - \frac{1}{2}\left(\frac{r_i - \theta_i}{\theta_i}\right)^2 + o_p\left(\frac{1}{n}\right),$$

where $o_p(f(n))$ denotes a term which is stochastically of lower order than $f(n)$, and we get

$$- 2n\ B(r;\theta) = n\ \sum_{i=1}^{k} \frac{(r_i - \theta_i)^2}{\theta_i} + o_p(1).$$

This is the Wilks' classical result which shows that the Pearson's chi-square test statistic is asymptotically equivalent to $-2n\ B(r;\theta)$, the log likelihood ratio statistic under the null hypothesis $\theta = \theta_0$, where θ_0 denotes the true distribution.

Since $B(r;\theta)$ is an estimate of $B(\theta_0;\theta)$ the present result answers the classical question why Pearson's chi-square can be useful when it is not apparently an estimate of anything [7, p.25].

The interpretation of the chi-square test of goodness of fit as a measurement of the entropy of the true distribution with respect to the assumed distribution can be extended to other commonly used chi-squares. Assume that a vector of real numbers $z = (z_1, z_2, \ldots, z_n)$ is given and we consider the use of the family of distributions $p(x|m,\sigma_0^2)$ $(-\infty < m < \infty)$ where $x = (x_1, x_2, \ldots, x_n)$ and x_i's are mutually independently distributed as $N(m,\sigma_0^2)$. As in the case of the multinomial family, the family of normal distributions is of exponential type and we have

$$\log\left[\frac{p(z|m,\sigma_0^2)}{p(z|m_*,\sigma_0^2)}\right] = B\{p(\cdot|m_*,\sigma_0^2);p(\cdot|m,\sigma_0^2)\},$$

where $m_* = (1/n) \Sigma z_i$, the maximum likelihood estimate of m. For a particular choice of $m = m_0$ we get

$$- 2\ B\{p(\cdot|m_*,\sigma_0^2);p(\cdot|m_0,\sigma_0^2)\} = \frac{n(m_* - m_0)^2}{\sigma_0^2} .$$

Thus the chi-square statistic on the right-hand side is twice the negentropy of the "true" model $p(\cdot|m_*,\sigma_0^2)$ with respect to $p(\cdot|m_0,\sigma^2)$.

If instead we consider the family of normal distributions $p(x|m,\sigma_0^2)$, where $m = (m_1, m_2, \ldots, m_n)$ with $-\infty < m_i < \infty$ and x_i's are mutually independent with x_i distributed as $N(m_i,\sigma_0^2)$, the maximum likelihood estimate of m is given by $m_* = z$ and we get

$$- 2\ B\{p(\cdot|m_*,\sigma_0^2);p(\cdot|m,\sigma_0^2)\} = \frac{\displaystyle\sum_{i=1}^{n} (z_i - m_i)^2}{\sigma_0^2} .$$

Under the assumption $m_i = m_0$ ($i = 1, 2, \ldots, n$) the above statistic takes the form

$$\frac{\displaystyle\sum_{i=1}^{n} (z_i - m_0)^2}{\sigma_0^2} .$$

This statistic is often suggested for the test of σ_0^2. However, our present derivation shows that it is twice the neg-entropy of $p(\cdot|z,\sigma_0^2)$ with respect to $p(\cdot|m_0,\sigma_0^2)$, where m_0 is identified with the n-vector $(m_0, m_0, \ldots, m_0)$. This observation suggests that the statistic would be more appropriate for the test of variability among the m_i's when σ_0^2 is known.

If we consider the family of distributions $p(x|m_0, \sigma^2)$
$(0 < \sigma^2 < \infty)$ with $m_i = m_0$ we get

$$- 2\, B\{p(\cdot|m_0, \sigma_*^2); p(\cdot|m_0, \sigma_0^2)\} = n\left(\frac{\sigma_*^2}{\sigma_0^2} - 1 - \log \frac{\sigma_*^2}{\sigma_0^2}\right),$$

where $\sigma_*^2 = (1/n)\, \Sigma(z_i - m_0)^2$. As a function of σ_0^2 this
quantity attains its minimum with $\sigma_0^2 = \sigma_*^2$ and will be useful
for the test of the hypothesis $\sigma^2 = \sigma_0^2$.

4. P-VALUE AND AIC

Berkson [9] discussed the evidential use of the test
of significance. This is concerned with the possible use of
P-values for the search of better models. Apparently there
is a tendency in practice to take a P-value as a standard
measure of the deviation of the hypothesis from the true
structure; for the discussion of this view, see for example,
Kempthorne [16] and Barndorff-Nielsen [8]. However, the
implication of the idea does not seem to have been pursued
to the point of practical consequence.

To clarify the essential point of the problem we consider the following simple example of polynomial fitting.
Assume that we are given a sequence of annual observations
$z(i)$ $(i = 1, 2, \ldots, 10)$. We also assume that the expansion
by normalized orthogonal polynomials is given by $z(i)$
$= a_0\phi_0(i) + a_1\phi_1(i) + \ldots + a_9\phi_9(i)$ $(i = 1, 2, \ldots, 10)$ with
$a_0 = 48.72$, $a_1 = 8.78$, $a_2 = 2.59$, $a_3 = -0.35$, $a_4 = 0.72$,
$a_5 = 0.68$, $a_6 = -1.10$, $a_7 = -0.31$, $a_8 = -0.39$ and $a_9 = -0.63$,
where $\phi_m(i)$ denotes the m-th order polynomial and it holds
that $\Sigma\phi_\ell(i)\phi_m(i) = 1$, for $\ell = m$, 0, otherwise. The
statistical model we assume is that $z(i)$'s are mutually
independently distributed as $N(f(i), 1)$ and the null hypothesis to be tested is the linearity of the regression $f(i)$.
If we restrict our family of distributions to those with k-th
order polynomial regression $(9 \geq k \geq 2)$, the log likelihood
ratio test statistic is given by

$$c_k = \sum_{m=2}^{k} a_m^2$$

which is to be tested against the chi-square distribution
with the degrees of freedom equal to k - 1. The problem
here is that there are several possibilities for the choice
of k.

That the problem is not trivial can be seen by the fact
that for the present data we get, for example,

$$\chi_1^2(0.01) < c_2(=6.70),$$

$$\chi_2^2(0.05) < c_3(=6.83) < \chi_2^2(0.01),$$

$$\chi_3^2(0.10) < c_4(=7.35) < \chi_3^2(0.05),$$

$$\chi_4^2(0.10) < c_5(=7.81) < \chi_4^2(0.05),$$

$$\text{and} \qquad c_6(=9.02) < \chi_5^2(0.10),$$

where $\chi_k^2(r)$ is defined by $\text{Prob}\{\chi_k^2(r) \leq \chi_k^2\} = r$ with the
chi-square variable χ_k^2 with the degrees of freedom k. This
result shows that the P-values of the chi-square tests
varies wildly depending on the choice of k. Particularly it
is rather puzzling to see the significance decreasing when k
is increased.

The chi-square variables c_k satisfies the relation

$$c_k = -2 \, B\{p_k(\cdot|_k\theta_*); p_1(\cdot|_1\theta_*)\},$$

where $p_k(\cdot|_k\theta)$ denotes the distribution of observations under
the assumption of the k-th polynomial regression, $_k\theta$ the
vector of the coefficients of regression and $_k\theta_*$ the maximum
likelihood estimate. By comparing the values of c_k's we are
comparing the deviations of $p_k(\cdot|_k\theta_*)$'s from $p_1(\cdot|_1\theta_*)$'s as
measured in terms of the entropies. Although the chi-square

variable c_k has this clear interpretation as twice the negentropy, for the purpose of comparison we have to pay our
attention to the possible differences of the statistical
variabilities of the chi-squares. For this we consider the
use of d_k = (-2) log P_k instead of c_k, where P_k
= Prob$\{c_k \leq \chi^2_{k-1}\}$. Under the assumption of the null hypothesis, P_k is uniformly distributed over [0,1] and d_k is distributed as a chi-square with the degrees of freedom equal
to 2. This means that by transforming c_k into d_k we are
normalizing the distributional characteristics of the
variables under an ideal condition.

To see the effect of such a transformation we calculated
e_k = c_k - d_k + 4 with c_k's put equal to $\chi^2_{k-1}(0.05)$ and
$\chi^2_{k-1}(0.01)$. The constant 4 was added to normalize the value
to 2 at k-1 = 1. Some values of e_k are given in Table 1.

Table 1. Values of e_k = c_k - d_k + 4

k - 1 (d.f.)	1	2	3	4	5	6
for c_k = $\chi^2_{k-1}(0.05)$	2.00	4.15	5.97	7.64	9.23	10.75
for c_k = $\chi^2_{k-1}(0.01)$	2.00	4.58	6.71	8.64	10.45	12.18

A remarkable point with these numerical values is that e_k's
are very close to 2(k-1) = 2 (degrees of freedom of the
chi-square). Thus for the purpose of comparison, at least
for these values of the degrees of freedom, we may replace
(-2) log P simply by chi-square - 2(degrees of freedom)
without invalidating the comparison of chi-squares at the
conventional critical level of 0.05 or 0.01.

However, we have the relation

$$c_k - 2(k-1) = AIC(1) - AIC(k),$$

where AIC(k) stands for an information criterion AIC of

the k-th order model defined by

AIC = (-2) log maximum likelihood
+ 2(number of estimated parameters).

The criterion AIC was introduced as an estimate of minus
twice the expected log likelihood of, or equivalently twice
the expected negentropy of the true distribution with respect
to, a model with the parameters determined by the method of
maximum likelihood [1]. Here we see the change of the status
of the k-th model from the "true" model which is used to
judge the significance of the null hypothesis (k = 1) to a
"null" model which is to be treated as an approximation to
the true distribution. Contrary to the chi-square, AIC is
defined without any reference to a particular null hypothesis
and is a measure of the badness of the model. The present
result shows that the comparison of (-2) log P_k's can be
justified if they are viewed as approximations to -AIC(k)'s.
    The values of AIC(k) = a_9^2 + a_8^2 + ... + a_{k+1}^2 + 2(k + 1)
are as follows: AIC(-1) = 2460.4, AIC(0) = 88.8, AIC(1)
= 13.7, AIC(2) = 9.0, AIC(3) = 10.8, AIC(4) = 12.3, AIC(5)
= 13.9, AIC(6) = 14.6, AIC(7) = 16.5, AIC(8) = 18.0 and
AIC(9) = 20.0. The minimum of AIC is attained by k = 2.
From the definition of AIC this means that the quadratic
regression is showing the best fit and is preferred to the
linear model. By switching from (-2) log P_k to AIC(k) we get
this clear-cut answer to the problem raised by Berkson.
    In the original definition of AIC only the correction
of the bias as an estimate of twice the negentropy was
considered. The present analysis revealed that it may also
be considered as distributionally normalized for a range of
applications of practical interest. (Unfortunately, there
is a tendency to identify AIC with an automatic model selection procedure realized by minimizing AIC, as represented
by the statement of Professor D. R. Cox in his presidential

address to the Royal Statistical Society (J. Roy. Statist.
Soc. Ser. A 144, (1981), p.293). AIC is only a measure of
the badness of a model and its definition stresses the
importance of a good modeling. The recent application of
AIC to clustering problems developed by H. Bozdogan (Doctoral
dissertation, University of Illinois at Chicago Circle, 1981)
is an example which demonstrates the potential of the criterion in data analytic situations.)

5. MODEL SELECTION AND BAYESIAN MODELING

Consider the use of a set of distributions $p_m(x)$ (m =
1, 2, ..., M) to realize an approximation to the distribution q(x). The approximation is realized by using data z.
A distribution $p(m|z)$ over m is generated as a function of
z and the $p_m(x)$ with m chosen by sampling from the distribution is adopted as an approximation of q(x). If z is obtained by sampling from a distribution $p_0(z)$, following [3],
the performance of the procedure is evaluated by the expected
entropy $E_z E_{m|z} B(q;p_m)$, or equivalently by
$E_z E_{m|z} E_x \log p_m(x)$, where E_z, $E_{m|z}$ and E_x denote the
expectations with respect to the distributions $p_0(z)$, $p(m|z)$
and q(x) respectively.

By Jensen's inequality we get the relation

$$E_z E_{m|z} E_x \log p_m(x) \le E_x \log E_m p_m(x),$$

where $E_m p_m(x) = E_z E_{m|z} p_m(x)$ is the mean of $p_m(x)$ with
respect to the distribution $p(m) = E_z p(m|z)$. This shows that
on the average $E_m p_m(x)$ or $\Sigma p_m(x) p(m)$ provides a better
approximation to q(x) than $p_m(x)$ specified by the random
sampling from $p(m|z)$. If we consider p(m) as the prior
probability of the model specified by $p_m(x)$ the present
result means that for any model selection procedure there is
a Bayesian model which provides a better approximation to
q(x) in the sense of maximizing the expected entropy.
Certainly the problem here is that in many applications $p_0(z)$
= q(z) and we cannot specify p(m).

Box [11] stressed the importance of predictive checking of a Bayesian model. In particular, he discussed the
test of a Bayesian model with an illustrative example.
The Bayesian model is defined with the data distribution
$p(x|m,\sigma^2)$, which is n-dimensional normal with the n components independently identically distributed as $N(m,\sigma^2)$ with
σ^2 known, and the prior distribution $p(m|m_0,\sigma_m^2)$, which is
one-dimensional normal with mean m_0 and variance σ_m^2. From
Box [11] the predictive distribution is defined by

$$p(x|m_0,\sigma^2,\sigma_m^2) = \int p(x|m,\sigma^2)\ p(m|m_0,\sigma_m^2)\ dm$$

and is given by

$$(-2)\ \log p(x|m_0,\sigma^2,\sigma_m^2) = n\ \log 2\pi + (n-1)\ \log \sigma^2$$
$$+ \log (\sigma^2 + n\sigma_m^2)$$
$$+ \frac{n}{\sigma^2}\ s_x^2 + \frac{n(\bar{x} - m_0)^2}{\sigma^2 + n\sigma_m^2}\ ,$$

where $\bar{x} - (1/n)\ \Sigma\ x_i$ and $s_x^2 - (1/n)\ \Sigma\ (x_i - \bar{x})^2$. When the
observation z is obtained, Box suggests the use of the
statistic

$$c_z = \frac{n}{\sigma^2}\ s_z^2 + \frac{n(\bar{z} - m_0)^2}{\sigma^2 + n\sigma_m^2}\ ,$$

which is a chi-square variable with the degrees of freedom
n, for the significance test of the Bayesian model.
Here we consider the family of Bayesian models defined
by $p(x|m,\sigma^2)$ and $p(m|\mu,\sigma_m^2)$, where $m = (m_1,\ m_2,\ \ldots,\ m_n)$,
$\mu = (\mu_1,\mu_2,\ \ldots,\ \mu_n)$ and the distribution $p(m|\mu,\sigma_m^2)$ is defined by the relations $m_i = \delta + \mu_i$ with δ distributed as $N(0,\sigma_m^2)$.
The corresponding predictive distribution $p(x|\mu,\sigma^2,\sigma_m^2)$ is
given by

$$(-2) \log p(x|\mu,\sigma^2,\sigma_m^2) = n \log 2\pi + (n - 1) \log \sigma^2$$
$$+ \log (\sigma^2 + n\sigma_m^2)$$
$$+ \frac{1}{\sigma^2} \sum_{i=1}^{n} (x_i - \mu_i - \bar{x} + \bar{\mu})^2$$
$$+ \frac{n(\bar{x} - \bar{\mu})^2}{\sigma^2 + n\sigma_m^2} ,$$

where $\bar{x} = (1/n) \sum x_i$ and $\bar{\mu} = (1/n) \sum \mu_i$. When the data z is obtained the maximum likelihood estimate of μ is given by $\mu_* = z$. We have the relation

$$c_z = (-2) \log p(z|m_0,\sigma^2,\sigma_m^2) + 2 \log p(z|\mu_*,\sigma^2,\sigma_m^2)$$

$$= -2 B\{p(\cdot|\mu_*,\sigma^2,\sigma_m^2) ; p(\cdot|m_0,\sigma^2,\sigma_m^2)\}.$$

This result shows that the chi-square test suggested by Box is measuring the deviation of the "true" distribution, obtained without putting any restriction on its means, from the assumed model. Thus the test will be useful for testing the assumption of the common mean m_0.

By comparing this example with the simple examples of section 3 we can see that there is no difference between the inference on the ordinary non-Bayesian model and that on the Bayesian model. The Bayes procedure which produces the posterior distribution is concerned only with the use of a given Bayesian model. The main aspect of inference lies entirely in the process of selecting a model. Here the only stochastic structure which completely specifies the distribution of the observation is the predictive distribution. Thus the measurement of the entropy of the predictive distribution becomes important. The difficulty of purely subjective justification of a Bayesian modeling is already discussed by Akaike [4].

The present example shows that the reduction of the expected variability of the fitted model is attained by the reduction of the number of unknown parameters by the adoption of the Bayesian model. This observation suggests that a criterion like AIC will be useful for the comparison of Bayesian models with the parameters determined by maximizing the likelihoods. Thus we can see that the gist of parametric modeling, either Bayesian or non-Bayesian, lies in the reduction of the number of parameters to be adjusted by data.

6. MEASUREMENT OF ENTROPY IN ACTION

In this section we will demonstrate by an example the dialectic interaction between the data and statistical model realized by the measurement of entropy. The example is concerned with the analysis of circadian rhythms of human physiology conducted by Dr Yutaka Honda of the Faculty of Medicine of the University of Tokyo. The original data are the measurements of saliva cortisol and body temperature taken nearly hourly, but with heavily missing observations, for 13 days, including a trip from Tokyo to Hawaii and back. To extract the daily pattern a modified version of the Bayesian seasonal adjustment program BAYSE [6] was applied to the data. This program performs the fitting of a Bayesian model defined by first representing the original observation of a time series x_i (i = 1, 2, ..., N) in the form

$$x_i = T_i + S_i + I_i,$$

where I_i is a normal white noise with mean zero and variance σ^2, and then assuming a prior distribution of the trend T_i and seasonal S_i. The prior distribution is defined by

$$p(T,S|\theta) \propto \exp\{-\frac{d^2}{2\sigma^2}(\alpha g(T) + h_1(S) + \beta h_2(S) + \gamma h_3(S))\},$$

where $g(T) = \Sigma (\Delta^k T_i)^2$, $h_1(S) = \Sigma (\Delta_p^\ell S_i)^2$, $h_2(S)$
$= \Sigma (S_i + S_{i-1} + \ldots + S_{i-p+1})^2$, $h_3(S) = \Sigma (\Delta^m S_i)^2$, where
$\Delta T_i = T_i - T_{i-1}$ and $\Delta_p S_i = S_i - S_{i-p}$, and θ denotes the
vector of necessary parameters. α, β and γ are positive
parameters. The inclusion of the term $h_3(S)$ represents the
modification of the original BAYSEA for the present
application.

The selection of the parameters is realized by minimizing the criterion ABIC which is defined as minus twice the
log likelihood of the Bayesian model; see [5] for details.
The criterion is an estimate of twice the negentropy which
measures the badness of the Bayesian model.

Fig.1 shows the original data of saliva cortisol and
the estimate, denoted as estimate A, of the circadian
rhythm obtained by a typical combination of the parameters.
By applying the same model to the record of body temperature
the result of Fig.2 was obtained. In contrast to the
estimate A of Fig.1 the estimated circadian rhythm of body
temperature shows a systematic variation due to the trip to
Hawaii. It lookes as if the stability of the 24 hour
circadian rhythm of saliva cortisol is confirmed. However,
by modifying a parameter the result denoted as estimate B
in Fig.3 was obtained. The criterion AVABIC, which is the
average of ABIC's of the Bayesian models fitted to successive
three-day spans of data, decreased from 891.9 of Fig.1 to
877.4 of Fig.3. Nevertheless the result of Fig.3 did not
look every informative compared with that of Fig.1. However,
a careful visual inspection of the estimate B of Fig.3
revealed the peculiar systematic behavior of the estimate
marked by arrows, thus suggesting the necessity of a check of
the original data. It was then found that the data was
exposed to some serious errors due to the misconception of a
research assistant who prepared the data for computation.

The estimate B of Fig.4 was obtained with the same parameters as those of the estimate B of Fig.3 but with the errors of the original data corrected. Similarly the estimate A of Fig.5 was obtained with the parameters of the estimate A of Fig.1. These results confirm the stability of the circadian rhythm of saliva cortisol. Although the estimate A of Fig.1 and that of Fig.5 look quite similar, the scientific contents of these results are obviously quite different.

Since the final conclusion is identical to that which would have been reached by assuming the validity of the original estimate A of Fig.1, the present result may be considered as a proof of the power of a Bayesian modeling when the validity of the model is established. The robustness of the procedure is obtained by throwing in the prior information through the specification of the prior distribution. Such application may be useful when the validity of the observations is questionable. In a scientific investigation of the sort of the present example, it is the validity of the model that is being checked. As is discussed in great depth by Box [11] these two phases of the interaction between the data and model take place at various levels of an investigation. The present example clearly demonstrates the importance of the measurement of entropies in a scientific investigation where progress takes place only through the generation of hypotheses.

ACKNOWLEDGEMENTS

The author is grateful to Dr Yutaka Honda for allowing him to use the preliminary results of the investigation of circadian rhythms. Thanks are due to Makio Ishiguro for the modification of BAYSEA and to Emiko Arahata for preparing the graphical outputs.

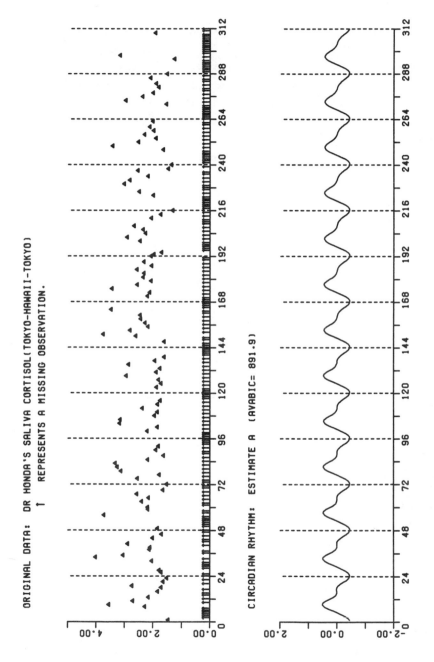

ORIGINAL DATA: DR HONDA'S SALIVA CORTISOL(TOKYO-HAWAII-TOKYO)
↑ REPRESENTS A MISSING OBSERVATION.

CIRCADIAN RHYTHM: ESTIMATE A (AVABIC= 891.9)

FIG.1. PRELIMINARY ESTIMATE OF CIRCADIAN RHYTHM OF SALIVA CORTISOL.

ORIGINAL DATA: DR HONDA'S BODY TEMPERATURE(TOKYO-HAWAII-TOKYO)
↑ REPRESENTS A MISSING OBSERVATION.

CIRCADIAN RHYTHM

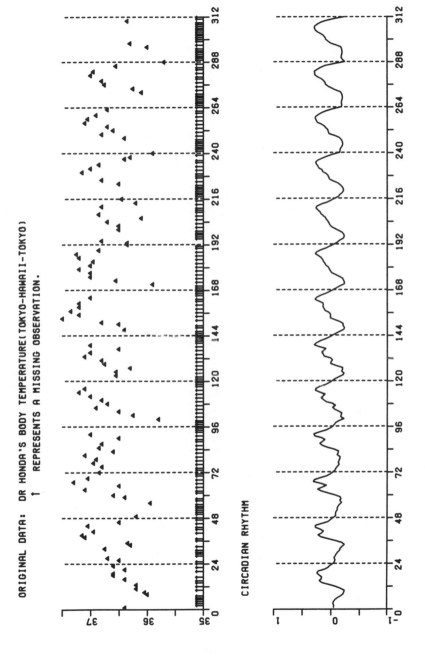

FIG.2. CIRCADIAN RHYTHM OF BODY TEMPERATURE.

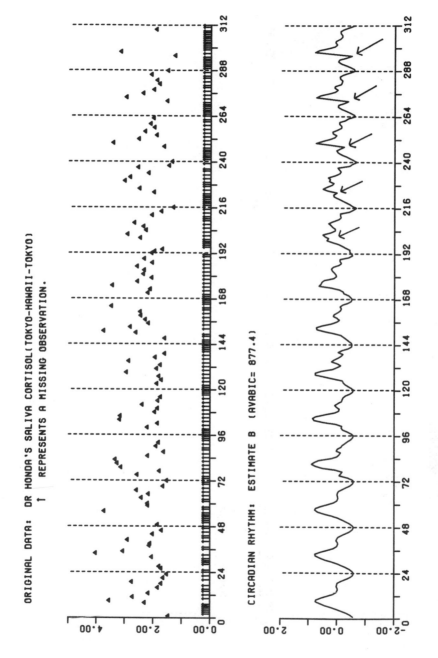

ORIGINAL DATA: DR HONDA'S SALIVA CORTISOL(TOKYO-HAWAII-TOKYO)
↑ REPRESENTS A MISSING OBSERVATION.

CIRCADIAN RHYTHM: ESTIMATE B (AVABIC= 877.4)

FIG.3. CIRCADIAN RHYTHM OF SALIVA CORTISOL (WITH LOWER AVABIC).

184

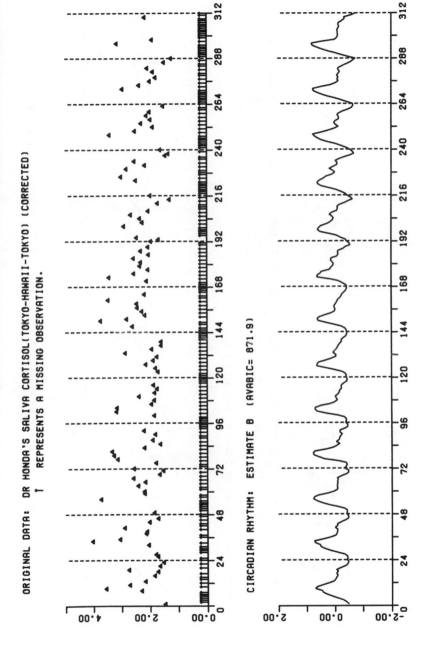

ORIGINAL DATA: DR HONDA'S SALIVA CORTISOL(TOKYO-HAWAII-TOKYO) (CORRECTED)
↑ REPRESENTS A MISSING OBSERVATION.

CIRCADIAN RHYTHM: ESTIMATE B (AVABIC= 871.9)

FIG.4. CIRCADIAN RHYTHM OF SALIVA CORTISOL (WITH CORRECTED DATA).

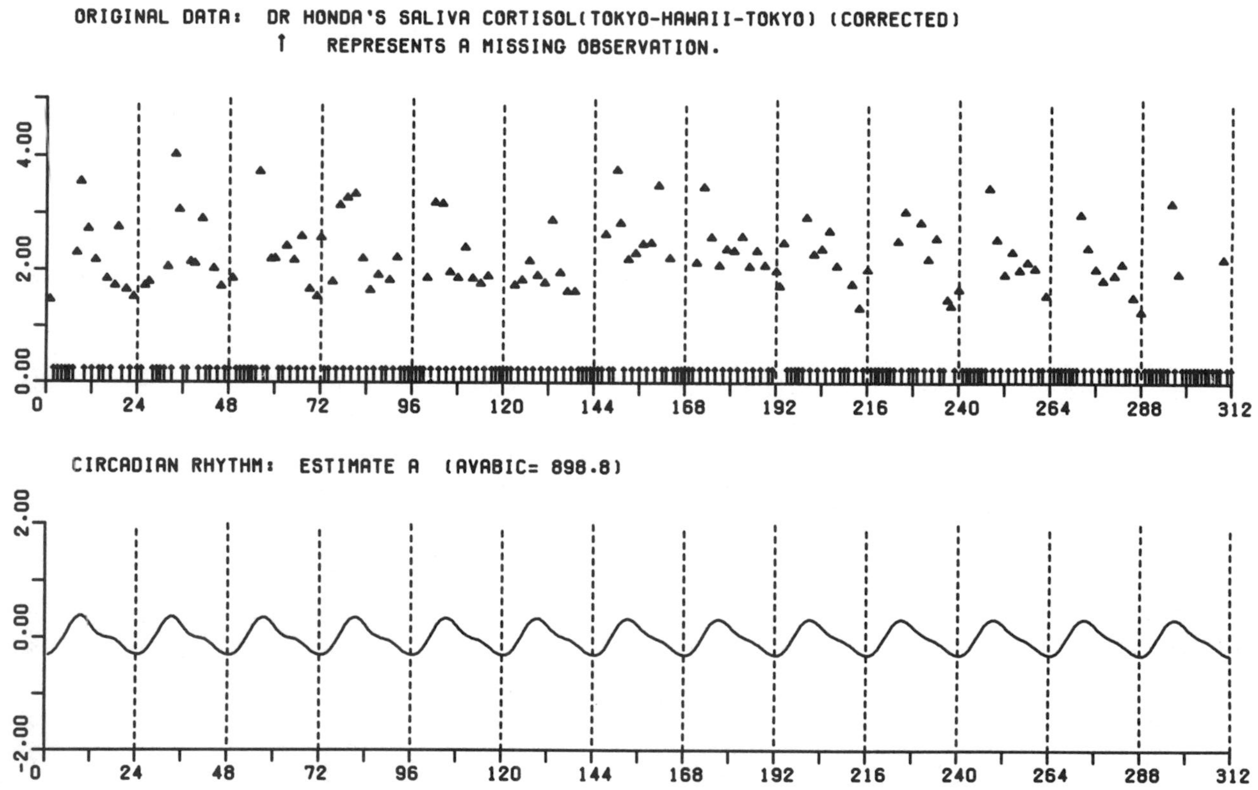
ORIGINAL DATA: DR HONDA'S SALIVA CORTISOL(TOKYO-HAWAII-TOKYO) (CORRECTED)
↑ REPRESENTS A MISSING OBSERVATION.
4.00
2.00
0.00
0
24
48
72
96
120
144
168
192
216
240
264
288
312
CIRCADIAN RHYTHM: ESTIMATE A (AVABIC= 898.8)
2.00
0.00
-2.00

FIG.5. CIRCADIAN RHYTHM OF SALIVA CORTISOL (WITH CORRECTED DATA).

REFERENCES

1. Akaike, H., A new look at the statistical model
    identification, IEEE Trans. Automat. Control AC - 19,
    (1974), 716 - 723.

2. _____, On entropy maximization principle, in
    Applications of Statistics (P. R. Krishnaiah, ed.),
    North - Holland, Amsterdam, 1977, 27 - 41.

3. _____, A new look at the Bayes procedure, Biometrika
    65, (1978), 53 - 59.

4. _____, Likelihood and the Bayes procedure, in
    Bayesian Statistics(J. M. Bernardo, M. H. DeGroot,
    D. V. Lindley and A. F. M. Smith, eds.), University
    Press, Valencia, Spain, 1980, 141 - 166.

5. _____, Seasonal adjustment by a Bayesian modeling,
    Journal of Time Series Analysis 1, (1980), 1 - 13.

6. _____ and Ishiguro, M., BAYSEA, A Bayesian Seasonal
    adjustment program, Computer Science Monographs No. 13,
    The Institute of Statistical Mathematics, Tokyo, 1980.

7. Anscombe, F., Contribution to the discussion of a paper
    by F. N. David and N. L. Johnson, J. Roy. Statist. Soc.
    Ser. B 18, (1956), 24 - 27.

8. Barndorff - Nielsen, O., Contribution to the discussion
    of a paper by D. R. Cox, Scand. J. Statist. 4, (1977),
    67 - 69.

9. Berkson, J., Tests of significance considered as evidence,
    J. Amer. Statist. Assoc. 37, (1942), 325 - 335.

10. Boltzmann, L., Uber die Beziehung zwischen dem zweiten
    Hauptsatze der mechanischen Warmetheorie und der
    Wahrscheinlichkeitsrechnung respektive den Satzen uber
    das Warmegleichgewicht, Wiener Berichte 76, (1877),
    73 - 435.

11. Box, G. E. P., Sampling and Bayes' inference in
    scientific modelling and robustness, J. Roy. Statist.
    Soc. Ser. A 143, (1980), 383 - 430.

12. Einstein, A., Uber einen die Erzeugung und Verwandlung
    des Lichtes betreffenden heuristischen Geschichtpunkt,
    Ann. Physik. 17, (1905), 132 - 148.

13. Fisher, R. A., On an absolute criterion for fitting
    frequency curves, The Messenger of Mathematics 41,
    (1912), 156 - 160.

14. _____, Statistical Method and Scientific
    Inference, Oliver and Boyd, Edinburgh, 1956.

15. Good, I. J., The contributions of Jeffreys to Bayesian
    statistics, in Bayesian Analysis in Econometrics and
    Statistics (A. Zellner, ed.), North - Holland,
    Amsterdam, 1980, 21 - 39.

16. Kempthorne, O., Theories of inference and data analysis,
    in Statistical Papers in Honour of G. W. Snedeccor
    (T. A. Bancroft, ed.), Iowa State University Press,
    Ames, Iowa, 1972, 167 - 191.

17. Klein, M. J., The development of Boltzmann's statistical
    ideas, Acta Phys. Austriaca, Suppl. 10, (1973),
    53 - 106.

18. Kullback, S., The two concepts of information, J. Amer.
    Statist. Assoc. 62, (1967), 685 -686.

19. Kupperman, M., Probabilities of hypotheses and
    information-statistics in sampling form exponential-
    class populations, Ann. Math. Statist. 29, (1958),
    571 - 575.

20. Pearson, K., On the criterion that a given system of
    deviations from the probable in the case of a correlated
    system of variables is such that it can be reasonably
    supposed to have arisen from random sampling,
    Philosophical Magazine (5) 50, (1900), 157 - 175.

21. Planck, M., Uber das Gesetz der Energieverteilung im
    Normalspectrum, Ann. Physik. 4, (1901), 553 - 563.

22. Ramsey, F. P., Truth and probability, (1926), in
    Studies in Subjective Probability (H. E. Kyburg, Jr.
    and H. E. Smokler, eds.), John Wiley, New York, 1964,
    61 - 92.

23. Simon, G., Additivity of information in exponential
    family probability laws, J. Amer. Statist. Assoc. 68,
    (1973), 478 - 482.

The Institute of Statistical
Mathematics
4-6-7 Minami-Azabu, Minato-ku
Tokyo 106, Japan

The Robustness
of a Hierarchical Model
for Multinomials and Contingency Tables

I. J. Good

1. INTRODUCTION: PHILOSOPHICAL MATTERS.

Jimmie Savage once said ironically that "philosophy" is a dirty ten-letter word. But since statistical associations are permissive societies I'll spend a few minutes on philosophical issues. Anyway others have been doing it at this conference, including George Box. His philosophy and mine are similar but I think not identical and some of the differences may be only terminological. After discussing a few philosophical points I'll exemplify them in terms of a Bayesian hierarchical model for categorical data. For further discussion of my philosophy see Good (1982).

The degree of belief in the approximate validity of a mathematical model, and its expected utility, depend on judgements and arguments both before and after observations are taken into account. This is true both for Bayesian and non-Bayesian models. (See, for example, Good, 1965, p. 40 and Box, 1981.) It is also true of all scientific theories. Some Bayesians say that in principle your priors can be fully determined by introspection, but "in principle" often means "impracticable" so to speak. One can't think of all possible hypotheses in advance. We have to be satisfied with models that are judged to be good enough, whether or not we know we are Bayesians. The people who don't know they are Bayesians are called *non-Bayesians*. This remark depends of course on how "Bayesian" is interpreted and I have counted 46,656 interpretations (Good, 1971b). This becomes 93,312 if we allow for the use or non-use of "dynamic probability" (Good, 1977). Dynamic probability is probability that changes by thought and calculation alone, without new empirical evidence. It is not at all a fancy idea and is essential for solving some philosophical

problems such as the one raised yesterday by Barnard from the floor, and
other similar ones that are even more interesting.

Suppose that a statistician rejects a hypothesis H on the grounds
that if H is true then something surprising has happened. The surprise
might be based on a tail-area probability or on a "surprise index"
(Weaver, 1948; Good, 1954, 1956b). The rejection of H by the statistician, even if the rejection is only provisional, requires that he believes
that the surprising event E, which might be the value of a statistic such
as χ^2, would appear less surprising given the non-null hypothesis $\bar{H}$, where
$\bar{H}$ can be as vague as you like. Now this belief depends on an implicit
judgement of $P(E|H)/P(E|\bar{H})$ or of $P(E|\bar{H})$, and this probability, which is
usually by no means sharp, is necessarily Bayesian, in a formal or informal sense, in the situation where $\bar{H}$ is composite, and usually $\bar{H}$ is highly
composite. Therefore "non-Bayesians" are Bayesians, at least informally.
In fact all people are Bayesians, because, for example, perception, recognition, and recall are Bayesian processes (consider, for example, the
delay in recognizing some one out of context: Good, 1950, p. 68) and
even non-Bayesians are people.

This argument is somewhat oversimplified because the usual reaction
when surprised is to look for new explanations, an undoubted fact of which
George Barnard reminded me at George Box's party in the evening after my
talk (compare Good, 1956b, p. 1131). The more surprising the event the
harder one looks for explanations, other things being equal, and if none
is found Barnard said he assumes that something remarkable happened by
chance. I would say that the more failure one has, in attempting to find
a simple alternative to H, the larger the probability that H is at least
a good approximation for the kinds of circumstances under which it has
been tested. But the probability does not tend to 1 because one might
have overlooked some explanations. I regard this as a Bayesian argument
that makes use of dynamic probability. It still seems to me therefore
that intelligent non-Bayesians are Bayesians. It may be that non-Bayesians simply don't use my terminology. For a discussion of allied matters,
see Good (1950, pp. 93-94; 1956a; 1980c; and 1981e).

I shall later refer to the use of tail-area probabilities in a Fisherian manner for the purpose of testing a Bayesian model and yet I have
just claimed that the use of tail-area probabilities depends on an informal Bayesian justification. This apparent circularity is one aspect of
the Bayes/non-Bayes compromise or synthesis (Good, 1957; 1965, Index).

A classical or sharp Bayesian selects a single prior for the ordinary parameters, but some modern Bayesians recognize some uncertainty in their priors. They try several priors, or a parametric family of priors. Parameters in priors are often called *hyperparameters*. A prior for a hyperparameter can be assumed and is called a *hyperprior*. Thus one arrives at a hierarchical Bayes procedure: see, for example, Good (1952, 1965, 1967, 1980b), Lindley (1971), Lindley & Smith (1972). A hierarchical model is sometimes called "two-stage" or "multistage" or "multilevel". I usually use the word "type", instead of "stage" or "level", because of the analogy with Russell's theory of types in mathematical logic.

Hierarchical Bayesian models can be used in a completely Bayesian manner or in a manner that has been called semi- or pseudo- or quasi-Bayesian. ("Quasi" sounds less pseudo than "pseudo".) In the latter case the hyperparameters, or possibly the hyperhyperparameters, etc., are estimated by non-Bayesian methods, or at least by some not purely Bayesian method such as by Maximum Likelihood. For an early example see Good (1956c), previously rejected in 1953. When a hyperparameter is estimated by Maximum Likelihood I prefer to call the method Type II Maximum Likelihood so as to make it clear that we are operating conceptually above the usual level though the level can be brought down as we shall see.

In so-called non-Bayesian statistics the use of the Ockham-Duns razor is sometimes called the principle of parsimony and it encourages one to avoid having more parameters than are necessary. This is one of the reasons for setting up null hypotheses and testing them. In hierarchical Bayesian methods one similarly uses a principle of parsimony or hyper-razor. To quote, "What can be done with fewer (hyper)parameters is done in vain with more" (Good, 1980b, p. 496). Once again it is possible to apply tests of one's model (Good, 1965, pp. 11, 40), and I shall give examples later.

There are at least two different ways to test a model. One is by means of significance tests after observations are made. Another is by examining the robustness of a model, that is, by seeing if small changes in the model lead to small changes in the implications. Tests for robustness can sometimes be carried out by the Device of Imaginary Results before making observations.

I shall exemplify these generalities in terms of a hierarchical Bayesian model for multinomials and contingency tables. For the sake of intelligibility it will be helpful to run through some of the history although much of it has already been given by Good (1965, 1980b).

2. MULTINOMIALS, FLATTENING CONSTANTS, AND MIXED DIRICHLET PRIORS.

Consider a multinomial population with t categories or cells and un-known physical probabilities $p_1, p_2, \ldots, p_t$ for the cells; and suppose we have a sample (n_i) $(i = 1, 2, \ldots, t)$, where $\Sigma n_i = N$, the sample size. The probabilities $p_1, p_2, \ldots, p_t$, or $t - 1$ of them, are the parameters of the population. Bayes and Laplace assumed a uniform distribution for n_1 or equivalently for p_1 for the binomial case $t = 2$ and arrived at the estimate $\hat{p}_1 = (n_1 + 1)/(N + 2)$ which is known as Laplace's Law of Succession. It is a shrinkage estimate. Bayes was the first shrink. Or maybe it was some one else in accordance with Stigler's Law that nothing is named after the originator. De Morgan in 1847 assumed a uniform distri-bution for the parameters of a multinomial in the simplex $\Sigma p_i = 1$ and arrived at the estimate $(n_i + 1)/(N + t)$ for p_i. He thus generalized Laplace's Law of Succession and was perhaps the first user of Dirichlet's multiple integral in statistics. For the De Morgan and some other refer-ences see Good (1980b or 1982). Bayes's billiard-table argument was ex-tended to multinomials by Good (1979).

The actuary G. F. Hardy used the beta distribution in 1889 as a prior for binomial estimation. This may have been the first use of a conjugate prior.

The philosopher W. E. Johnson (1932) showed that if the estimate of p_1 does not depend on the ratios of $n_2 : n_3 : \ldots : n_t$ (his sufficientness assumption), then this estimate must be of the form $(n_1 + k)/(N + tk)$ for some positive k, so that k can be regarded as a flattening constant. His proof was correct for $t > 2$. As pointed out by Good (1965, p. 25), Johnson's formula is precisely equivalent to the assumption of a symmetric Dirichlet prior density

$$D(k, t) = \Gamma(tk)\Gamma(k)^{-t} \Pi p_i^{k-1} \quad (k > 0). \tag{1}$$

(Note that we shall throughout ignore any information that there might be in the *order* of the t categories.) Similarly the use of t flattening con-stants, that is, taking $(n_i + k_i)/(N + \Sigma k_i)$ as the posterior logical prob-ability of the ith cell, is equivalent to the use of the general Dirichlet prior density

$$D((k_i)) = \Gamma(\Sigma k_i)\Pi p_i^{k_i-1}/\Pi\Gamma(k_i) \quad (k_i > 0; i = 1, 2, \ldots, t) \tag{2}$$

which is the usual generalization of the beta density. The lumping prop-erty of the Dirichlet distribution when categories are combined follows at once from the interpretation of the k_i's as flattening constants.

Various estimates of p_i correspond to symmetric Dirichlet priors. De Morgan's estimate corresponds to $k = 1$, Jeffreys's invariant prior to $k = \frac{1}{2}$, Perks's invariant prior to $k = 1/t$, Maximum Likelihood (ML) estimation to $k = 0$, though it is not necessary to interpret the ML estimate n_i/N as a logical or subjective probability. It would be an absurd betting probability if $n_i = 0$ or $n_i = N$ because it would lead to offering odds of your soul to a centime.

In Good (1965) I argued that it is better to treat k as a hyperparameter rather than as a constant, at least when t is large. [Note the misprint on p. 46 where $(2k)^{\frac{1}{2}}$ should be $(2k)^{-\frac{1}{2}}$.] For a completely Bayesian approach one can assume a hyperprior density $\phi(k)$, which is equivalent to using an ordinary prior density

$$\int_0^\infty D(k, t)\phi(k)\, dk, \tag{3}$$

or one can use a quasi-Bayesian approach in which k is estimated by Type II Maximum Likelihood. I did not claim perfection, but it seemed to me that a mixture of Dirichlet priors was a better reasonably general-purpose prior for multinomials than the individual Dirichlet priors considered previously. (The replacement of Dirichlet *processes* by *mixtures* was advocated by Good, 1978b.)

One application of this approach is for testing the hypothesis of equi-probability

$$H_\infty: \; p_1 = p_2 = \ldots = p_t = 1/t,$$

the class of alternatives being $\{H_k\}$ where H_k asserts that the prior density is $D(k, t)$. The Bayes factor in favor of H_k (the factor by which its odds are multiplied) as compared with H_∞, provided by the evidence (n_i), is

$$\begin{aligned} F(k) &= F(H_k/H_\infty : (n_i)) \\ &= t^N \Gamma(tk) \Pi_i \Gamma(n_i + k)/\Gamma(k)^t \Gamma(N + tk) \\ &= \Pi_{i=1}^{t}\, \Pi_{j=1}^{n_i-1}\, (1 + j/k)/\Pi_{j=1}^{N-1}\, \{1 + (j/tk)\} \end{aligned} \tag{4}$$

and

$$F = F(\cup_k H_k/H_\infty : (n_i)\,|\,\phi) = \int_0^\infty F(k)\phi(k)\,dk. \tag{5}$$

(The colon is pronounced "provided by".) The elementary form of (4) can be proved without using Dirichlet integrals. The hyperprior $\phi(k)$ has to be proper or at least $\int^\infty \phi(k)\,dk$ must converge (at $k = \infty$), otherwise $F = 1$, that is, the evidence would be wiped out. The reason for this is that

$F(k) \to 1$ as $k \to \infty$. (See Fig. 1.) In particular, we must not take $\phi(k) = 1/k$ which is the Jeffreys-Haldane uninformative prior density for positive parameters. This density can, however, be approximated, over a wide range of values of k, by a log-Cauchy density

$$\phi(k) = \phi(k, \lambda, \mu) = \frac{1}{\pi k} \cdot \frac{\lambda}{\lambda^2 + [\log_e (k/\mu)]^2} \qquad (6)$$

in which λ and μ are hyperhyperparameters. Note that μ is the median of the hyperprior while the upper and lower quartiles q and q' are given by $q = \mu e^{\lambda}$ and $q' = \mu e^{-\lambda}$. If you can guess or judge the upper and lower

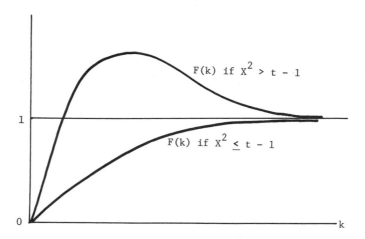

FIG. 1. THE GRAPH OF $F(k)$ IF NO $n_i = N$. Similarly, for contingency tables, if $F_3(k)$ is unimodal, as seems highly probable, then it has the same general shapes unless the number ν of empty cells is the maximum ν permitted by the marginal totals; and the condition for the upper shape is $R > E(R)$. If $\nu = \nu_0$, then $F_3(0)$ is positive. (By assumption, no marginal total vanishes.)

quartiles of the flattening constant k you can compute λ and μ from the equations

$$\lambda = \tfrac{1}{2} \log (q/q'), \quad \mu = (q q')^{\frac{1}{2}}. \qquad (7)$$

In this model the subjective input occurs in the choice of the hyperhyperparameters. You can check your judgements in terms of other quantiles: the nth percentile is (Good, 1969, p. 46)

$$\mu(q/\mu)^{-\cot(\pi n/100)}. \qquad (8)$$

Inference seems not to depend much on the choice of q and q' over a wide range; that is, we seem to have Bayesian robustness in this respect. An example will be given later in this paper.

The actuary Wilfrid Perks (1946) wanted the prior to be invariant under the lumping together of categories and proposed therefore that the flattening constant should be inversely proportional to the number of categories. Since, for $t = 2$ he wanted $k = \frac{1}{2}$, in accordance with his invariance theory (Perks, 1947; cf Jeffreys, 1946 who had $k = \frac{1}{2}$ for all t), he therefore wanted $k = 1/t$ in general. By means of a numerical example with $N = t = 100$ I convinced him that this prior was unsatisfactory (Good, 1967), after failing to do so in an interchange of 37 philosophical letters. This was another example of the Device of Imaginary Results.

We shall see presently that there is a way of achieving what Perks was trying to achieve but "at a higher level". But first let us consider the testing of the general null hypothesis

$$H_{\infty,\pi} : p_1 = \pi_1, \ p_2 = \pi_2, \ \cdots, \ p_t = \pi_t.$$

Given the non-null hypothesis $\overline{H}_\pi$, we could use the prior densities

$$D(k, \ t, \ \pi) = \frac{\Gamma(kt)}{\Pi\Gamma(kt\pi_i)} \ \Pi \, p_i^{kt\pi_i - 1} \tag{9}$$

and

$$\int_0^\infty D(k, \ t, \ \pi)\phi(k)dk \tag{10}$$

(Good, 1967, pp. 408-409. There are two misprints in formula (25) of that paper.)

Under this generalization formula (4) is replaced, in a self-explanatory notation, by

$$\begin{aligned}
F(k, \ \pi) &= F(H_{k,\pi}/H_{\infty,\pi} \ : \ (n_i)) \\
&= \frac{\Pi\pi_i^{-n_i} \Gamma(tk)\Pi\Gamma(n_i + kt\pi_i)}{\Pi\Gamma(kt\pi_i)\Gamma(N + tk)} \\
&= \Pi_{i=1}^t \Pi_{j=1}^{n_i-1}(1 + j/kt\pi_i)/\Pi_{j=1}^{N-1}(1 + j/kt)
\end{aligned} \tag{11}$$

and (5) is applicable with $F(k)$ replaced by $F(k, \pi)$. As a function of k, $F(k, \pi)$ has at most one local maximum, as proved by Levin & Reeds (1974). See Fig. 1.

The use of the prior (9) is equivalent to using flattening constants $kt\pi_i$ that are proportional to the "widths" of t subintervals on the unit interval, so that it can be regarded as a generalization of Perks's use of the flattening constant inversely proportional to the number of categories. But I prefer to use the mixture (10) of Dirichlet prior densities, and the invariance under lumping of categories can be achieved provided that $\phi(k)$ is of the form $\psi(kt)/t$. For a special case of this result see the discussion below related to contingency tables.

To prove this invariance property, let us take the density function (10) with $\phi(k) = \psi(kt)/t$ and write it out explicitly as

$$\int_0^\infty \frac{\Gamma(kt)}{\Pi\Gamma(kt\pi_i)} \, \Pi \, p_i^{kt\pi_i-1} \, \frac{\psi(tk)}{k} \, dk.$$

Imagine the last two cells lumped together and write $t' = t - 1$, $\pi_i' = \pi_i$ ($i \leqslant t - 2$), $\pi_{i'}' = \pi_{t-1} + \pi_t$, with a similar notation for p_i' ($i = 1, 2, \ldots, t'$). By the lumping property of the Dirichlet distribution the density becomes

$$\int_0^\infty \frac{\Gamma(kt)}{\Pi_{i=1}^{t'}\Gamma(k+\pi_i')} \, \Pi_{i=1}^{t'} \, p_i'^{kt\pi_i'-1} \, \frac{\psi(tk)}{k} \, dk.$$

Let's change to a new variable $k' = kt/t'$, then the density is seen to be

$$\int_0^\infty D(k', \, t', \, \pi_i', \, \ldots, \, \pi_{t'}') \, \frac{\psi(t'k')}{k'} \, dk'$$

which is of the form (10) with t decreased to t'. The asserted invariance property now follows by an obvious application of mathematical induction. We have thus achieved Perks's aim but one level up the hierarchy.

Note that the mixing function $\psi(tk)/k$ *can equally be used for continuous Dirichlet processes*, a comment that supplements Good (1974).

I'll say more about the choice of $\phi(.)$ later.

3. CONTINGENCY TABLES.

The work was extended to contingency tables by Good (1965, 1976) and Crook & Good (1980). The latter paper deals with the question of robustness under variations in the hyperhyperparameters but it does not give numerical results for the very important case of 2×2 tables. Some examples for 2×2 tables will be given here.

There are various methods for sampling contingency tables. Let's consider the following three familiar methods, called methods 1, 2, and 3. In method 1 the sample is chosen at random from the whole population (multinomial sampling), in method 2 the row totals are fixed, and in method 3 both the row and column totals are fixed. The corresponding Bayes factors against the hypothesis H of "independence" (no association) are called F_1, F_2, and F_3 respectively. The Bayes factor against H provided by the row and column totals alone, when the sampling is by method 1, is equal to F_1/F_3, and we call this FRACT. (See Good, 1976, and Crook

& Good, 1980.) The corresponding Bayes factor provided by the row totals alone is F_1/F_2, but we consider that this is equal to 1 in most circumstances. We half agree with Fisher so to speak, because he assumed that FRACT = 1, though of course that was not the way he expressed the matter. We predict that most statisticians will agree that the row totals alone usually contribute entirely negligible evidence in favor of association.

When we sample by method 1, the row totals of an $r \times s$ contingency table (n_{ij}) with sample size N have a multinomial distribution and we assume their prior density is $D^*(r, 1)$ for the sake of consistency with the multinomial model where

$$D^*(t, u) = \int_0^\infty D(t, uk)\phi(k)dk. \qquad (12)$$

Similarly we assume the prior density $D^*(s, 1)$ for the column totals, and we regard these two priors as independent given H. Similarly, given $\bar{H}$ we assume the prior $D^*(rs, 1)$ for the cell probabilities. This implies that the priors for the row totals and column totals, given $\bar{H}$, are respectively $D^*(r, s)$ and $D^*(s, r)$. Note that, given $\bar{H}$, we are treating the rs cells of the table as a multinomial. This is the simplest procedure but there are many other possible procedures especially when r and s are large.

The assumption that $F_1 = F_2$, made by Crook & Good (1980), implies a constraint on the Bayesian model, namely that $\phi(k)$ must be of the form $\psi(tk)/k$. This was first noted by Good, 1978a, published by Good, 1980a, and full details were given by Crook & Good, 1980. Note that this gave information about the hyperprior for multinomials because any multinomial *could* be regarded as arising as the row totals of a contingency table. The further extension at the end of Section 2 is new.

The condition on $\phi(.)$ was met by Crook & Good (1980) by taking

$$\phi(k) = \phi(k, \lambda, \mu) = \frac{\lambda}{k\pi\{\lambda^2 + [\log_e(k/\mu)]^2\}} \qquad (13)$$

usually with $\lambda = \pi$, $\mu = 1/t$. (In the context of 2×2 tables t is always either 2 or 4.) Thus the resultant prior for the row totals, given H, is

$$\int_0^\infty D(r, k)\phi(k, \pi, 1/r)dk. \qquad (14)$$

The prior for the row totals, given $\bar{H}$, is

$$\int_0^\infty D(r, sk)\phi(k, \pi, 1/(rs))dk = \frac{1}{s}\int_0^\infty D(r, k)\phi(\frac{k}{s}, \pi, \frac{1}{rs})dk$$

which is the same as (14). Thus the priors have the property that the row totals alone, and the column totals alone, provide no evidence concerning H.

By using the formulae for F_1 and F_3 given by Good (1976, p. 1165) we find that

$$\text{FRACT} = F_1/F_3 = \frac{\Sigma^* \phi((m_{ij}), rs)}{\phi((n_{i.}),r)\phi((n_{.j}),s)} \tag{15}$$

where the row totals are $(n_{i.})$, the column totals are $(n_{.j})$,

$$\phi((\ell_\nu), t) = \frac{(\Sigma_1^t \ell_\nu)!}{\Pi_1^t (\ell_\nu!)} \int_0^\infty \frac{\Gamma(tk)\Pi\Gamma(\ell_\nu + k)}{[\Gamma(k)]^t \Gamma(N + tk)} \phi(k, \pi, \frac{1}{t}) \, dk, \tag{16}$$

and where the summation Σ^* in the numerator of (15) is over all contingency tables (m_{ij}) having the assigned marginal totals $(n_{i.})$ and $(n_{.j})$ $(i = 1, 2, \ldots, r;\ j = 1, 2, \ldots, s)$. When $N = 0$ or 1, we have FRACT $= 1$ which is a check.

4. NUMERICAL RESULTS FOR CONTINGENCY TABLES, AND ROBUSTNESS

Numerical values of FRACT were given by Crook & Good (1980), but the only 2×2 tables considered there had all four marginal totals equal to say n. It was then found that FRACT was very closely approximated by $0.028(\log_2 n)^2 + 0.0071(\log_2 n) + 2.172$ for $2 \leqslant n \leqslant 1024$ when the hyperprior (13) was assumed. Table 1 herein gives some more examples for 2×2 contingency tables. The figures were computed by Dr. J. F. Crook and are extracted from a more extensive table, which is qualitatively

TABLE 1. FRACT, THE BAYES FACTOR AGAINST "INDEPENDENCE" PROVIDED BY ROW AND COLUMN TOTALS IN 2×2 CONTINGENCY TABLES. The values of FRACT are tabulated for $\lambda = \frac{1}{2}\pi$ and π, and $\mu = 1/t$ and $16/t$. For example, when the row totals are $(3, 17)$, the column totals are $(6, 14)$, and $\lambda = \pi$, then the values of FRACT are 1.3 and 1.2 when $\mu = 1/t$ and $16/t$ respectively.

Column totals	Row totals	(3, 17)	(6, 14)	(9, 11)	$(\lambda, t\mu)$	q	q´
(3, 17)		6.9	1.3	.65	$(\pi, 1)$	11.6	.022
		6.6	1.2	.58	$(\pi, 16)$	185	.35
		4.7	1.1	.68	$(\frac{1}{2}\pi, 1)$	2.4	.10
		4.4	1.1	.62	$(\frac{1}{2}\pi, 16)$	38	1.7
(6, 14)		1.3	2.6	1.6	$(\pi, 1)$	11.6	.022
		1.2	1.6	1.2	$(\pi, 16)$	185	.35
		1.1	2.7	1.5	$(\frac{1}{2}\pi, 1)$	2.4	.10
		1.1	1.4	1.1	$(\frac{1}{2}\pi, 16)$	38	1.7
(9, 11)		.65	1.6	2.3	$(\pi, 1)$	11.8	.022
		.58	1.2	1.5	$(\pi, 16)$	185	.35
		.68	1.5	2.6	$(\frac{1}{2}\pi, 1)$	2.4	.10
		.62	1.1	1.3	$(\frac{1}{2}\pi, 16)$	38	1.7

similar, in an unpublished joint note dated 1980 September 29. *The results apply to both sampling methods 1 and 2.* To see the effect of vary-
ing the hyperpriors we have taken $\lambda = \frac{1}{2}\pi$ and π, and $\mu = 1/t$ and $16/t$.
Thus, for the binomial margins, the subjective lower quartiles of the
flattening constant vary from $\frac{1}{2}e^{-\pi} = .02$ to $8e^{-\frac{1}{2}\pi} = 1.7$ and the upper
quartiles from $\frac{1}{2}e^{\frac{1}{2}\pi} = 2.4$ to $8e^{\pi} = 185$. In view of these very wide
spreads for each quartile, we feel justified in saying that we have
Bayesian robustness. This will be especially true if, when using sam-
plingmethod 2, the row totals (or the column totals) are taken to be
equal, which is often natural to do.

The results are consistent with what Crook & Good (1980) found, and
explained intuitively, for larger contingency tables, namely that FRACT
is largest when both row and column marginal totals are very rough, and
is smallest when one side is very rough and the other if flat. In practice most contingency tables sampled by methods 1 or 2 are between these
extremes so that FRACT is usually between 1/2 and 5/2. Therefore, when
testing independence, it doesn't make much difference which sampling
model is used.

The entries on the diagonal of Table 1 are somewhat larger than
their neighbours. The intuitive reason for this is that, for example,
when the row and column margins of a 2 × 2 table are both (6, 14), it is
possible to have a table (6, 0; 0, 14), which would convey much evidence
in favor of association.

It is interesting that Fisher's judgement is somewhat supported by
our Bayesian model, in the sense that FRACT is not usually very far from
1, though of course the model does not support his somewhat extreme position of regarding the evidence from the combined marginal totals as entirely negligible.

When calculating a Bayes factor F_1 reasonable robustness with respect to the choice of hyperhyperparameters has also been found (Crook
& Good, 1980, p. 1202).

5. THE TYPE II LIKELIHOOD RATIO.

One aspect of the Bayes/non-Bayes compromise or synthesis is that a
Bayes factor can be used as a statistic for significance tests. That is,
its distribution can be examined in a Fisherian spirit (Good, 1957, p.
863; 1965, p. 35). I believe that this will often be a good method
even when the Bayesian model is not very good. The method has been used
for the Bayesian models described above, which models are I think quite

good. A statistic G was defined in terms of a *type II likelihood ratio*
for testing H_∞ within the composite hypothesis $\cup_k H_k$ and its distribution
was found, under the null hypothesis, to be approximately proportional
to a standard normal distribution, the approximation being good down to
incredibly small tail-area probabilities (Good & Crook, 1974). The definition of G for multinomials is

$$G = \{2 \log_e \max_k F(k)\}^{\frac{1}{2}}. \tag{17}$$

The condition for $G \neq 0$ (i.e. $G > 0$) is $X^2 > t - 1$, where

$$X^2 = (t/N) \Sigma (n_i - N/t)^2. \tag{18}$$

[In Good, 1967, p. 411, in the line after (37), G^2 was misprinted as G.]
Similarly, the definition of G for contingency tables is

$$G = \{2 \log_e \max_k F_3(k)\}^{\frac{1}{2}}, \tag{19}$$

where the notation $F_3(k)$ is self-explanatory. Here the precise condi-
tion for $G > 0$ is (Good, 1981b) that $R > E(R)$, where $R = \Sigma \frac{1}{2} n_{ij}(n_{ij} - 1)$
is the number of repeats within the cells of the contingency table, and

$$E(R) = \Sigma \binom{n_i \cdot}{2} \Sigma \binom{n_{\cdot j}}{2} / \binom{N}{2}. \tag{20}$$

This, as a necessary and sufficient condition, depends on the conjecture
that $F_3(k)$ is unimodal, which is probably both true and very difficult
to prove. The condition $X^2 > E(X^2)$, which I gave in Crook & Good (1980),
is not quite correct: it was the wrong generalization of the multinomial
condition.

Methods for calculating G approximately are given by Good (1981a, c),
but improved versions of both these notes are in the works.

6. WHEN CAN ONE REJECT THE SYMMETRICAL DIRICHLET PRIORS FOR MULTINOMIALS?

As mentioned in the Introduction, when we are not sure of our priors
it seems appropriate to some of us to use the observations to test the
priors. Let us apply the Neyman-Pearson-Wilks likelihood ratio test, one
level up (Type II Likelihood Ratio), to test the prior D(k, t), that is,
the hypothesis H_k, and also to test $\cup_k H_k$. We can test these hypotheses
within the wider class D(**k**, t). Note first the predictive probability
(see, for example, Good, 1965, p. 36, or Levin & Reeds, 1974), the multi-
nomial-Dirichlet probability:

$$P((n_i) | (k_i)) = \frac{N! \, \Pi \, [k_i(k_i+1)\ldots(k_i+n_i-1)]}{[\Pi \, n_i!] \, K(K+1)\ldots(K+N-1)} \quad (K = \Sigma k_i) \tag{21}$$

which tends to the maximum possible value

$$\frac{N!}{\Pi n_i!} \, \Pi \, (\frac{n_i}{N})^{n_i} \tag{22}$$

when $k_i/n_i \to \infty$ for each i. Therefore testing within $D(\underline{k}, t)$ by the Type II likelihood ratio test is the same as testing within all possible multinomials. Now

$$P((n_i)|k) = \frac{N! \, \Pi \, [k(k+1)\dots(k+n_i-1)]}{[\Pi n_i!] \, tk(tk+1)\dots(tk+N-1)} \tag{23}$$

so the Type II Likelihood Ratio criterion is

$$\Lambda(k) = \Lambda(k : (n_i)) = 2\{-N \log_e N + \Sigma \, n_i \log_e n_i$$
$$- \Sigma[\log_e k + \log_e(k+1) + \dots + \log_e(k+n_i-1)]$$
$$+ \log_e(tk) + \log_e(tk+1) + \dots + \log_e(tk+N-1)\} \tag{24}$$

and this has asymptotically a χ^2 distribution with t degrees of freedom if H_k is true. Likewise $\Lambda(k_{max} : (n_i))$ has asymptotically a χ^2 distribution with t - 1 degrees of freedom if $\cup_k H_k$ is true, where k_{max} is the value of k that maximizes (24). This requires that $X^2 > t - 1$. [To maximize (24) look for a zero of its derivative by using the Newton-Raphson method starting the iteration with a value of k appreciably less than $N/(X^2 - t + 1)$.]

Examples of these criteria are given in Table 2. It was not, of course, obtained by simulation. For every sample (n_i) in Table 2 we have $\Lambda(k_{max}) < X^2$ and this may well be invariably true. For very rough samples $\Lambda(k_{max})$ is much smaller than X^2. When the sample (n_i) is not too far from equiprobable, as measured by X^2 or by its tail-area probability $P(X^2)$, we see that the prior $\cup_k D(k, t)$ is acceptable. The specific values $k = 1$, $k = \frac{1}{2}$ and $k = 1/t$ are much less reliable and, for these values of k, we often have $\Lambda(k) > X^2$. For samples (n_i) that are extremely rough the symmetric Dirichlet priors can be "rejected" both individually and as a class, but in such cases it would still be all right to use the priors for rejecting H_∞ because H_∞ would be obviously false. This comment is an example of "marginalism" (Good, 1958, pp. 808-809; 1969, p. 61; 1971a, p. 15). Putting it briefly: When testing a null hypothesis H within a composite hypothesis $\overline{H}$, it is only in marginal cases that it is important to use an accurate prior for the components of $\overline{H}$ (conditional on $\overline{H}$).

It might be possible to make effective use of a prior of the form

$$\int_0^\infty \dots \int_0^\infty D((k_i)) \phi(k_1, k_2, \dots, k_t) dk_1 \dots dk_t \tag{25}$$

TABLE 2. VALUES OF $\Lambda(k) = \Lambda(k : (n_i))$ DEFINED BY (24), TOGETHER WITH THEIR P VALUES. $\Lambda(k_{max})$ has $t - 1$ degrees of freedom, but $\Lambda(k)$ has t degrees of freedom.

t	n_1	n_2	n_3	n_4	n_5	n_6	n_7	n_8	n_9	n_{10}	N	x^2	k_{max}	$\Lambda(k_{max})$	P	$\Lambda(1)$	P	$\Lambda(\tfrac{1}{2})$	P	$\Lambda(\tfrac{1}{t})$	P
2	1	19	–	–	–	–	–	–	–	–	20	16.2	.41	3.51	.061	4.14	.126	3.54	.17	3.54	.17
2	2	18	–	–	–	–	–	–	–	–	20	12.8	.65	3.46	.063	3.58	.167	3.50	.17	3.50	.17
2	5	15	–	–	–	–	–	–	–	–	20	5.0	2.32	2.63	.105	2.89	.236	3.48	.17	3.48	.17
2	1	99	–	–	–	–	–	–	–	–	100	96	.23	4.49	.034	7.24	.027	5.14	.076	5.14	.076
2	10	90	–	–	–	–	–	–	–	–	100	64	.65	5.02	.025	5.18	.075	5.06	.080	5.06	.080
2	30	70	–	–	–	–	–	–	–	–	100	16	3.27	3.77	.052	4.34	.114	5.06	.080	5.06	.080
2	40	60	–	–	–	–	–	–	–	–	100	4	16.5	2.39	.122	4.21	.122	5.06	.080	5.06	.080
5	15	4	17	6	8	–	–	–	–	–	50	13.0	4.22	8.73	.068	11.0	.051	13.6	.018	18.3	.0026
5	6	8	18	5	13	–	–	–	–	–	50	11.8	5.41	8.02	.091	10.9	.053	13.6	.018	18.4	.0025
5	8	8	8	8	18	–	–	–	–	–	50	8.0	13.1	5.9	.207	10.6	.060	13.6	.018	18.5	.0024
5	3	3	3	3	8	–	–	–	–	–	20	5.0	21.3	4.1	.393	7.48	.187	10.1	.072	14.8	.011
10	1	1	1	1	1	1	1	1	1	51	60	375	.40	26.2	.0019	31.7	.00045	26.4	.0032	35.2	.00012
10	0	0	0	0	0	0	0	1	1	58	60	301	.056	16.4	.059	45.6	<10^{-5}	30.7	.00066	17.0	.074
10	5	5	5	5	5	5	5	5	5	55	100	225	1.16	27.3	.0012	27.4	.0023	30.3	.00076	47.8	<10^{-5}
10	7	7	7	7	7	7	7	6	7	37	100	81	2.80	21.5	.011	24.9	.0055	30.3	.00076	49.9	<10^{-5}
10	5	15	4	2	13	1	18	6	17	19	100	45	1.78	25.3	.0026	26.4	.0032	30.3	.00076	48.5	<10^{-5}
10	19	8	6	15	2	13	15	5	12	5	100	28	3.98	20.3	.016	24.8	.0057	30.3	.00076	49.9	<10^{-5}
10	7	13	6	4	11	3	16	8	15	7	100	23.4	5.34	18.5	.030	24.4	.0066	30.3	.00076	50.3	<10^{-5}
10	15	7	7	7	10	14	17	3	14	6	100	19.8	7.46	16.7	.054	24.2	.0071	30.3	.00076	50.5	<10^{-5}
10	5	15	6	18	11	8	7	13	11	6	100	17.0	11.6	14.4	.109	23.8	.0081	30.3	.00076	50.8	<10^{-5}

to avoid having to invoke marginalism, but, as mentioned by Good (1967, p. 409), this would probably involve too much calculation unless t is small. Moreover Table 2 suggests that that generality of (25) is unnecessary.

If we are testing the hypothesis $p_i = \pi_i$ (i = 1, 2, ... , t) by using the Dirichlet priors D(k, t, $\underline{\pi}$) and their mixture (10), then presumably the results just stated will extend in an obvious manner. (24) is replaced by

$$(D(k, t, \underline{\pi}) : (n_i)) = 2\{-N \log_e N + \Sigma n_i \log_e n_i$$
$$- \Sigma [\log_e(kt\pi_i) + \ldots + \log_e(kt\pi_i + n_i - 1)]$$
$$+ \log_e(tk) + \ldots + \log_e(tk + N - 1)\}. \quad (24*)$$

The sum here is empty for i for which n_i vanishes.

It is at first surprising that $\Lambda(\frac{1}{2} : (n_i))$ depends almost entirely on N and t alone. It is explained by means of Stirling's formula which leads to

$$\Lambda(\tfrac{1}{2} : (n_i)) \sim \log_e \{\tfrac{1}{2}(N + \tfrac{1}{2}t)^{2N+t-1} N^{-2N} t^{-t+1}\}, \quad (26)$$

provided that $n_i \neq 0$ (i = 1, 2, ... , t).

7. SCIENTIFIC INDUCTION, UNIVERSAL AND PREDICTIVE.

Scientific induction is concerned with at least two problems apart from the *formulation* of hypotheses: (i) is a hypothesis always true ("universal" induction, to use a hypallactic epithet)? and (ii) will it succeed on the next trial or the next M trials (predictive induction)? In the multinomial context we have say $n_1 = N$ and the first inductive problem is whether $p_1 = 1$. Call this hypothesis J_1. For some literature on induction see, for example, Keynes (1921, Index), Jeffreys (1961, p. 43), and Good (1972, 1981d).

In our model the Bayes factor in favor of J_1 provided by the observation $n_1 = N$ is approximately the reciprocal of

$$P(n_1 = N | \upsilon_k H_k) = \int_0^\infty \frac{\Gamma(k+N)\Gamma(tk)}{\Gamma(k)\Gamma(tk+N)} \phi(k) dk, \quad (27)$$

by (23). Of course in general the larger t is the smaller the initial odds of J_1. If N >> t this integral is approximately

$$\frac{1}{t} \int_0^\infty N^{-(t-1)k} \phi(k) dk$$

and this depends mainly on the behavior of $\phi(k)$ where $k = O(1/(t \log N))$. For general forms of the hyperprior the integral therefore cannot be robust, but for the log-Cauchy form it is fairly robust as one can see from Table 3. Of course the posterior odds of J_1, besides being inversely pro-

TABLE 3. THE FINAL ODDS OF THE HYPOTHESIS J_1 THAT $p_1 = 1$. Assuming that the initial odds of J_1 are $\beta/(t - 1)$, the entries in the table give the approximate final odds divided by β. The entries are obtained by dividing $(t - 1)$ into the reciprocal of the integral (27) with ϕ defined by (6).

(i) t = 2

N = 2	10	20	50	500	10^6	10^{12}	λ	$t\mu$	q	q´
2.8	4.4	4.8	5.1	5.6	6.7	7.8	π	1	11.6	.022
3.4	7.9	8.6	9.1	9.9	11.4	12.7	π	16	185	.35
2.6	4.2	4.7	5.3	7.3	9.9	12.5	$\frac{1}{2}\pi$	1	2.4	.10
3.2	9.6	10.8	11.5	13.7	15.6	17.2	$\frac{1}{2}\pi$	16	38	1.7

(i) t = 10

N = 2	10	20	50	500	10^6	10^{12}	λ	$t\mu$	q	q´
2.1	2.9	3.1	3.4	3.6	4.3	4.8	π	1	11.6	.022
3.2	5.2	5.6	5.8	6.2	6.9	7.9	π	16	185	.35
1.9	3.1	3.4	3.7	5.3	6.9	8.6	$\frac{1}{2}\pi$	1	2.4	.10
5.3	7.4	7.6	7.9	8.6	9.3	10.1	$\frac{1}{2}\pi$	16	38	1.7

portional to the integral, are proportional to the prior odds which might vary widely from one scientist to another, and the ratio of the upper and lower odds as estimated by a single scientist might be large. Largely because of the variability of the prior odds, universal induction is not very robust. It will often be reasonable to assume that the prior odds are roughly of the form $\beta(t - 1)$, where β is interval-valued. For example, the prior *probability* that all ravens are the same color might well be inversely proportional to the number of colors because the more finely we define color the less likely is the hypothesis. Under these assumptions the final odds will be roughly

$$\beta[\int_0^\infty N^{-x} \psi(x)\frac{dx}{x}]^{-1} \tag{28}$$

which is mathematically independent of t. For example, if the first hundred ravens are jet black, the probability that all ravens are jet black is about the same as if the adjective "jet" has been omitted in both places.

Table 3 has been set out to allow for the assumption that the prior odds of J_1 are of the form $\beta(t - 1)$. The table confirms that the change from t = 2 to t = 10 then makes little difference to the final odds of J_1. Also, for a fixed value of β, the final odds of J_1 do not vary much when the interval (q´, q) is varied a lot so that within our model universal induction would be robust if the initial odds were uncontroversial. A feature of Table 3 worth special attention is that the odds of J_1 increase very slowly as N is increased and would be large only if N were more than astronomically large, say if $N = \exp(10^{10})$.

The other form of scientific induction is concerned with the probability that the next item sampled will belong to category 1 given that $n_1 = N$, where N is large. This probability is

$$p + (1 - p)\int_0^\infty \frac{N+k}{N+tk} \, \phi(k)\,dk \tag{29}$$

where p is the prior probability of J_1. This formula ignores the negligible contributions arising from the prior probabilities of special values of p_1 other than 1 such as 0, $\frac{1}{2}$, and computable values in the sense of Turing (Good, 1950, p. 55n). Formula (29) is approximately

$$p + (1 - p)(1 - a/N) = 1 - a(1 - p)/N, \tag{30}$$

where

$$a = \frac{(t - 1)\mu}{\pi} \int_0^\infty \frac{e^{-\lambda y}\,dy}{1 + y^2}$$

$$= -\frac{(t - 1)\mu}{\pi} [(\sin \lambda)\,ci(\lambda) + (\cos \lambda)\,si(\lambda)] \tag{31}$$

and is therefore proportional to μ, the prior median of k. The expression in brackets is $-.28$ when $\lambda = \pi$ and is $-.47$ when $\lambda = \frac{1}{2}\pi$. (31) reduces to

$$\frac{(t-1)\mu}{\pi} \times .28 \quad (\text{if } \lambda = \pi); \quad \frac{(t-1)\mu}{\pi} \times .47 \quad (\text{if } \lambda = \frac{1}{2}\pi).$$

The probability that the next M observations are all of the first category (when the first N are) is approximately

$$p + (1 - p)(1 - \frac{a}{N})(1 - \frac{a}{N+1}) \cdots (1 - \frac{a}{N+M})$$

$$\approx p + (1 - p)(1 + M/N)^{-a}. \tag{32}$$

If $M = o(N)$, the probability is approximately

$$p + (1 - p)e^{-Ma/N} \approx 1 - aM(1 - p)/N. \tag{33}$$

The probability (of a "success") on the next trial does not depend much on the prior probability p of the universal hypothesis J_1 and it also does not depend much on the values of the hyperparameters.

Thus predictive induction is more robust than universal induction.

I am indebted to Drs. J. F. Crook and L.-F. Lee for the programming.

REFERENCES[†]

1. Box, G.E.P. (1980). "Sampling and Bayes' inference in scientific modelling and robustness", *J. Roy. Statist. Soc. Ser. A, 143*, 383-430 (with discussion).

2. Crook, J. F. & Good, I. J. (1980). "On the application of symmetric Dirichlet distributions and their mixtures to contingency tables, Part II", *Ann. Statist. 8* (Nov.), 1198-1218.

3. Good, I. J. (1950). *Probability and the Weighing of Evidence* (London, Charles Griffin; New York, Hafners; pp. 119).

4. *----- (1952). "Rational decisions", *J. Roy. Statist. Soc. Ser. B, 14*, 107-114.

5. *----- (1954). "The appropriate mathematical tools for describing and measuring uncertainty". *Uncertainty and Business Decisions* (ed. C.F. Carter, G.P. Meredity, and G.L.S. Shackle; Liverpool: University Press, 1954; 2nd. edn. 1957), pp. 20-36.

6. ----- (1956a). Contribution to the discussion of a paper by G.S. Brown. In *Information Theory: Third London Symposium* (Colin Cherry, ed.), p. 13. London: Butterworths.

7. ----- (1956b). "The surprise index for the multivariate normal distribution", *Ann. Math. Statist. 27*, 1130-1135; *28* (1957), 1055.

8. ----- (1956c). "On the estimation of small frequencies in contingency tables", *J. Roy. Statist. Soc. Ser. B, 18*, 113-124.

9. ----- (1957). "Saddle-point methods for the multinomial distribution", *Ann. Math. Statist. 28*, 861-881.

10. ----- (1958). "Significance tests in parallel and in series", *J. Amer. Statist. Assoc. 53*, 799-813.

11. ----- (1965). *The Estimation of Probabilities: An Essay on Modern Bayesian Methods*, MIT Press, pp. xii + 109.

12. ----- (1967). "A Bayesian significance test for multinomial distributions", *J. Roy. Statist. Soc. Ser. B, 29*, 399-431 (with discussion).

13. *----- (1969). "A subjective analysis of Bode's law and an 'objective' test for approximate numerical rationality", *J. Amer. Statist. Assoc. 64*, 23-66 (with discussion).

14. ----- (1971a). Contribution to the discussion of a paper by J. Neyman. In *Foundations of Statistical Inference* (V.P. Godambe & D.A. Sprott, eds.: Toronto: Holt, Rinehart & Winston), pp. 14-15.

[†]The items marked with an asterisk are to appear, sometimes with ellipses, in Good, 1982).

15.   ----- (1971b). "46656 varieties of Bayesians", letter in *Amer. Stat. 25* (Dec.), 62-63.

16.   ----- (1972). "Scientific induction and exponential-entropy distributions", *Amer. Stat. 26*, p. 45.

17.   ----- (1976). "On the application of symmetric Dirichlet distributions and their mixtures to contingency tables", *Ann. Statist. 4*, 1159-1189.

18.   *----- (1977). "Dynamic probability, computer chess, and the measurement of knowledge", in *Machine Intelligence 8* (eds. E.W. Elcock and D. Michie; Ellis Horwood Ltd. & Wylie), 139-150.

19.   ----- (1978a). "The information in the marginal totals of a contingency table", in the Special Lecture Series on "R.A. Fisher: an appreciation" at the University of Minnesota, May 23.

20.   ----- (1978b). Review of Ferguson, Thomas S., "Prior distributions on spaces of probability measures", Ann. Statist. *2*, 615-629; *MR 55*, pp. 1546-1547, Rev. #11479.

21.   ----- (1979). "Bayes's billiard-table argument extended to multinomials", C44 in *J. Statist. Comput. Simulation 9*, No. 2, 161-163.

22.   ----- (1980a). "The contributions of Jeffreys to Bayesian statistics", in *Bayesian Analysis in Econometrics and Statistics: Essays in Honor of Harold Jeffreys* (ed. Arnold Zellner; Amsterdam: North Holland), Chap. 3, pp. 21-34.

23.   *----- (1980b). "Some history of the hierarchical Bayesian methodology", *Trabajos de Estadistica ye de Investigacion Operativa*. Also in *Bayesian Statistics: Proceedings of the First International Meeting held in Valencia (Spain), May 28 to June 2, 1979* (J.M. Bernardo, M.H. DeGroot, D.V. Lindley, and A.F.M. Smith, eds.; Univ. of Valencia, 1981), 489-510 & 512-519 (with discussion).

24.   ----- (1980c). "The diminishing significance of a P-value as the sample size increases", C73 in *J. Statist. Comput. Simulation 11*, Nos. 3 & 4, 307-309.

25.   ----- (1981a). "An approximation of value in the Bayesian analysis of contingency tables", C88 in *J. Statist. Comput. Simulation 12*, No. 2, 145-147.

26.   ----- (1981b). "When is G positive in the mixed Dirichlet approach to contingency tables?" C94 in *J. Statist. Comput Simulation 13*, 49-52.

27,   ----- (1981c). "The Monte Carlo computation of Bayes factors for
       contingency tables", C95 in *J. Statist. Comput. Simulation 13*,
       52-56.

28.   ----- (1981d). "Can scientific induction be meaningfully ques-
       tioned?" C100 in *J. Statist. Comput. Simulation 13*, 154.

29.   ----- (1981e). "Surprise, surprise", being a contribution to the
       discussion of a paper by G. Shafer. *J. Amer. Statist. Assoc.*

30.   ----- (1982). *Good Thinking: The Foundations of Probability and
       its Applications* (University of Minnesota Press).

31.   Good, I. J. & Crook, J. F. (1974). "The Bayes/non-Bayes compromise
       and the multinomial distribution", *J. Amer. Statist. Assoc. 69*
       (Sept.), 711-720.

32.   Jeffreys, H. (1946). "An invariant form for the prior probability
       in estimation problems", *Proc. Roy. Soc. London Ser. A, 186*, 453-
       461.

33.   Jeffreys, H. (1961). *Theory of Probability*, 3rd edn. Oxford: Uni-
       versity Press.

34.   Johnson, N.L. & Kotz, S. (1969). *Discrete Distributions* (Boston:
       Houghton-Miflin).

35.   Johnson, W.E. (1932). Appendix (ed. by R.B. Braithwaite) to "Prob-
       ability: deductive and inductive problems", *Mind 41*, 421-423.

36.   Keynes, J.M. (1921). *A Treatise on Probability*. London & New
       York: Macmillan.

37.   Levin, B. & Reeds, J. (1974). "Maximum likelihood estimation of
       'flattening constants' in multinomial-Dirichlet distributions and
       and a proof of a conjecture of I.J. Good", Research Report S-27,
       Dept. of Statistics, Harvard University, pp. 18. (A shorter ver-
       sion in *Ann. Statist. 5*, 1977, 79-87.)

38.   Lindley, D.V. (1971). "The estimation of many parameters". In
       Foundations of Statistical Inference (V.P. Godambe & D.A. Sprott,
       eds.; Toronto: Holt, Rinehart & Winston), 435-455.

39.   Lindley, D.V. & Smith, A.F.M. (1972). "Bayes estimates for the
       linear model", *J. Roy. Statist. Soc. Ser. B, 34*, 1-41 (with
       discussion).

40.   Perks, W. (1947). "Some observations on inverse probability in-
       cluding a new indifference rule", *J. Inst. Actuaries 73*, 285-312
       (with discussion).

41.   Weaver, W. (1948). "Probability, interest, rarity, and surprise",
       Sci. Monthly 67, 390-392.

The work was supported by an N.I.H. grant #GM18770.

Department of Statistics
Virginia Polytechnic Institute and
State University
Blacksburg, Virginia 24061

A Case Study of the Robustness
of Bayesian Methods of Inference:
Estimating the Total in a Finite Population
Using Transformations to Normality

Donald B. Rubin

1. PROLOGUE-THE PRACTICAL INTERPRETATION OF INTERVAL
   ESTIMATES AS BAYES INTERVALS

Bayesian methods of inference will be, I believe, the
primary statistical tools used to analyze data in the
future, at least in those cases in which the purpose of
statistical analysis is to provide a range of likely values
for an unknown quantity, such as the total in a finite
population or the relative effect of a treatment in an
experiment. One reason for this belief is the inherent
flexibility of Bayesian models with their multiple levels
of randomness; such methods naturally lead to smoothed
estimates in complicated data structures and consequently
possess the ability to obtain better real world answers.

Another reason for this belief that Bayesian methods
will constitute the standard tools for providing interval
estimates is more psychological, and involves the
relationship between the statistician and the client who is
the consumer of the statistician's work. In nearly all
practical cases, clients will interpret intervals provided
by statisticians as Bayesian intervals, that is, as
probability statements about the likely values of unknown
quantities conditional on the evidence in the data. Such
direct probability statements require prior probability
specifications for unknown quantities, and thus the kinds of
answers clients will assume are being provided by

statisticians, Bayesian answers, require prior probability
assumptions. If the Bayesian answers vary dramatically for
different reasonable assumptions unassailable by the data,
then the resultant range of Bayesian answers must be
entertained as legitimate, and I believe that the
statistician has the responsibility to make the client aware
of this fact.

Of course, there are assumptionless confidence intervals, but these are not generally useful inferentially. For
an extreme example, consider the following 95% confidence
interval: regardless of the values of the data, 95% of the
time the interval is $(-\infty, \infty)$ and 5% of the time the
interval is $[0,0]$. Confidence intervals are generally
useful and fair summaries of data only when they can be
interpreted as approximate (or, in some circumstances,
conservative) Bayesian intervals.

In brief, interval estimates will be interpreted by
clients as Bayesian (or approximately Bayesian) intervals
and therefore statisticians have an obligation to try to
provide interval estimates that can legitimately be
interpreted as such, or at least to offer guidance as to
when the intervals that are provided can be safely
interpreted in this manner.

2.   THE ROBUSTNESS OF BAYESIAN METHODS

The potential application of statistical methods is
often demonstrated either (a) theoretically, (b) from
artificial data generated following some convenient analytic
form, or (c) from real data without a known correct
answer. But quite generally, we understand tools through
the consequences of their application, and these three kinds
of demonstrations, although useful, provide somewhat limited
evidence on how well the tools can be expected to work in
practice. The case study presented here uses a small, real
data set with a known value for the quantity to be
estimated. It is surprising and instructive to see the care
that may be needed to arrive at satisfactory inferences with
real data.

The specific example concerns the estimation of the total population of the N = 804 municipalities in New York State from a simple random sample of n = 100 (source = Encyclopedia Britannica, 1960 census; New York City was represented by its five boroughs). Table 1 presents summary statistics for this population and two simple random samples. These two samples were the first and only ones chosen. With knowledge of the population, neither sample appears particularly atypical; sample 1 is very representative of the population, whereas sample 2 has a few too many large values. Consequently, it might at first glance seem straightforward to estimate the population total, perhaps overestimating the total from the second sample.

This example was originally studied to demonstrate the relative ease with which Bayesian models could be fit to such data using simulation techniques to approximate posterior distributions, and the example does illustrate this point. It does not, however, generate the message that these techniques can be automatically applied to arrive at sound inferences. Rather, it dramatizes three important messages.

The first two messages are concrete and address the accuracy of resultant inferences for covering the true population total.

(1) Although the log normal model is often used to estimate the total on the raw scale (e.g., estimate total pollutant, medical costs or oil reserves assuming the logarithm of the values are normally distributed), the log normal model may not provide accurate inferences for the total even when it appears to fit fairly well as judged from probability plots.

(2) Extending the log normal family to a larger family, such as the Box-Cox family of power transformations, and selecting a better fitting model by Bayesian/likelihood criteria or probability plots may lead to less realistic inferences for the population total, even when probability plots indicate an adequate fit.

TABLE 1:   Summary Statistics for Populations of
            Municipalities in New York State; All
            804 and Two Simple Random Samples of 100
            (Source: Encyclopedia Brittanica -
            1960 Census Figures)

	Population N = 804	Sample 1 N = 100	Sample 2 N = 100
Total	13,776,663	1,966,745	3,850,502
Mean	17,135	19,667	38,505
Std. Dev.	139,147	142,218	228,625
Low	19	164	162
5%	336	308.5	315
25%	800	891.5	863
Med.	1,668.5	2,081.5	1,740
75%	5,050	6,049.5	5,239
95%	30,295	25,130	41,718
High	2,627,319	1,424,815	1,809,578

These two points are not criticisms of the log transformation or the Box-Cox family of power transformations.
Rather, they are warnings about the naive statement "better
fits to data mean better models which in turn mean better
real world answers". Statistical answers rely on prior
assumptions as well as data, and better real world answers
generally require models that incorporate more realistic
prior assumptions as well as provide better fits to data.
This comment naturally leads to the last message of this
paper, which is a general one encompassing the first two.

(3) In general, inferences are sensitive to features
of the underlying distribution of values in the population
that cannot be addressed by the observed data.
Consequently, for good statistical answers we need

(a) models that allow observed data to dominate
prior restrictions,

and either

(b) flexibility in these models to allow
specification of realistic underlying features of
population values not adequately addressed by
observed values, such as behavior in the extreme
tails of the distribution,

or

(c) questions that are robust for the type of
data collected in the sense that all relevant
underlying features of population values are
adequately addressed by the observed values.

Finding models that satisfy 3a and 3b is a more general
approach than finding questions that satisfy 3c because
statisticians are often presented with hard questions that
require answers of some sort, and do not have the luxury of
posing easy (i.e. robust) questions in their place. For
example, for environmental reasons it may be important to
estimate the total amount of pollutant being emitted by a
manufacturing plant using samples of the soil from the
surrounding geographical area, or, for purposes of budgeting
a health-care insurance program, it may be necessary to

estimate the total amount of medical expenses from a sample
of patients. Such questions are inherently nonrobust in
that their answers depend on the behavior in the extreme
tails of the underlying distributions. Estimating more
robust population characteristics, such as the median amount
of pollutant in soil samples or the median medical expense
for patients, does not address the essential questions in
such examples.

At least from a Bayesian perspective, the more major
effort in statistics currently seems to be focused on 3c
rather than on 3a and 3b, that is on defining the estimand
to be the midmean or the population analogue of some other
robust estimator of location. Although such work is
obviously important, it seems somewhat surprising that less
effort is being devoted to the development of
computationally attractive tools that are capable of
addressing both easy and hard questions, especially since
the current collection of statistical tools satisfying both
criteria 3a and 3b seems to be rather limited.

This third point is not a criticism of any particular
tool for inference, but it is a criticism of the claim that
inferential tools, such as the jackknife (c.f. Miller, 1974)
or bootstrap (Efron, 1980, Rubin, 1981) can be assumption
free. We need to define conditions (i.e., prior assumptions, data, and questions) under which a particular
statistical tool works well and those conditions under which
it does not. Moreover, we must cautiously interpret statements like "normal looking samples automatically provide
robust estimates of location" and "if it can't be estimated
well, it won't affect inferences" as well as "if the data do
not contradict the model, the model is satisfactory for
drawing inferences". All statements are true under
particular conditions but generally are false: in general,
inferences depend on assumptions that the data at hand
cannot address. Robustness of Bayesian inference is a joint
property of data, prior knowledge, and questions under

consideration; the remainder of this article illustrates this general point in our example.

3. SAMPLE 1 -- INITIAL ANALYSIS

We begin the data analysis by trying to estimate the population total from Sample 1. The standard 95% interval for the finite population total is:

$$N\bar{y} \pm 2 \, s \, N \sqrt{\frac{1}{n} - \frac{1}{N}} \, . \tag{1}$$

For our problem $N = 804$, $n = 100$, and for Sample 1, the sample mean, $\bar{y}$, equals 19,667 and the sample standard deviation, s, equals 142,218. Hence, the observed value of interval (1) is approximately

$$(-5.6 \times 10^6, \, 37.2 \times 10^6) \, . \tag{2}$$

Interval (2) can be justified under certain assumptions as a 95% interval from either the randomization theory perspective (c.f. Cochran, 1963) or the Bayesian perspective (c.f. Ericson, 1969; Rubin, 1978). From either perspective, the required assumptions are not well supported with a skew sample like Sample 1, but are supported with approximately normally distributed samples.

The practical man examining the standard 95% interval (2) might find the upper limit useful and simply replace the lower limit by the total in this sample, since the total in the population can be no less; this procedure would give a 95% interval estimate of $(2 \times 10^6, \, 37 \times 10^6)$ for the population total. We note that this does cover the true population total, 14×10^6.

Surely, modestly intelligent use of statistical models should produce a better answer because from Table 1, both the population and Sample 1 are very far from normal, and the standard interval is most appropriate with normal populations. Even before seeing any data, we know that sizes of municipalities are far more likely to look something like log normal than normal. Figures 1 and 2 show normal and log-normal probability plots for Sample 1.

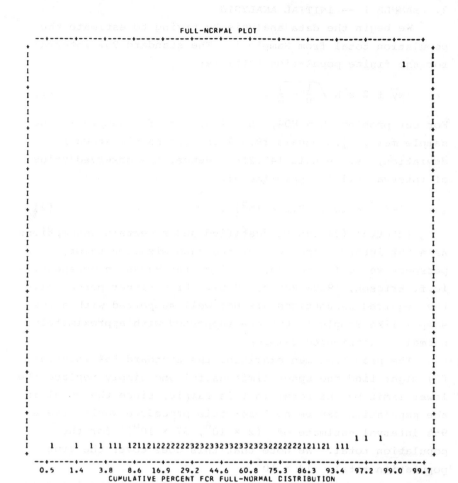

Figure 1: Normal Plot, Y_i, Sample 1.

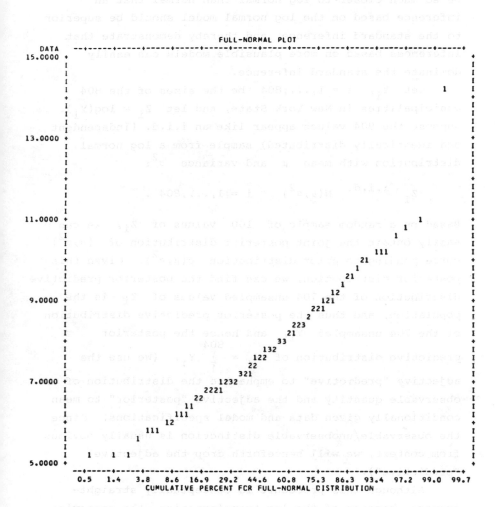

Figure 2: Normal Plot, Log(Y_i), Sample 1.

Although the data do not appear to be exactly log normal
(primarily because of one extreme value), they do appear to
be so much closer to log normal than normal that an
inference based on the log normal model should be superior
to the standard inference, and thereby demonstrate that
inferences based on more plausible models can easily
dominate the standard inference.

Let Y_i, $i = 1,\ldots,804$ be the sizes of the 804
municipalities in New York State, and let $Z_i = \log(Y_i)$.
Suppose the 804 values appear like an i.i.d. (independent
and identically distributed) sample from a log normal
distribution with mean μ and variance σ^2:

$$Z_i \overset{i.i.d.}{} N(\mu,\sigma^2) \qquad i = 1,\ldots,804 \ .$$

Based on a random sample of 100 values of Z_i, we can
easily obtain the joint posterior distribution of (μ,σ^2)
corresponding to prior distribution $p(\mu,\sigma^2)$. Given this
posterior distribution, we can find the posterior predictive
distribution of the 704 unsampled values of Z_i in the
population, and thus the posterior predictive distribution
of the 704 unsampled Y_i, and hence the posterior
predictive distribution of $Y_+ = \sum_1^{804} Y_i$. (We use the
adjective "predictive" to emphasize the distribution of an
observable quantity and the adjective "posterior" to mean
conditionally given data and model specifications. Since
the observable/unobservable distinction is usually obvious
from context, we will henceforth drop the adjective
"predictive").

Although this procedure is conceptually straightforward, because of the log transformation, the posterior
distribution for Y_+ cannot be written in simple closed
form. Consequently, we will approximate the posterior
distribution of Y_+ using simple simulation techniques.
The Appendix outlines the simulation procedure. With prior
distribution $p(\mu,\sigma^2) \propto \sigma^{-1}$, the posterior distribution of
μ given σ^2 is $N(\overline{Z},\sigma^2)$ and the posterior distribution

of σ^2 is s_z^2 times an inverted χ^2 on 99 d.f. Consequently, it is easy to draw (μ, σ^2) from its posterior
distribution. Having drawn values of μ and σ^2, say μ_*
and σ_*^2, it is easy to draw 804 values from the posterior
distribution of Z_i, i = 1,804, given $\mu = \mu_*$ and
$\sigma^2 = \sigma_*^2$: values of Z_i that are in the sample are fixed
at their observed values and the 704 unsampled values of
Z_i are drawn as i.i.d. $N(\mu_*, \sigma_*^2)$. Summing the 100 observed
values of $Y_i = \exp(Z_i)$ and the 704 drawn values of
$Y_i = \exp(Z_i)$ gives one value of Y_+ drawn from its
posterior distribution. Note that any other feature of the
population, such as the 95th percentile, can be calculated
at this time. Drawing a second value of (μ, σ^2) and
repeating the process yields a second value of Y_+.

We drew 100 values of Y_+ which are displayed in Stem-
and-Leaf 1. Based on these 100 simulated values, we find
that the posterior median of Y_+ is approximately
6.9×10^6, and the 95% interval based on the third and
97th of the 100 drawn values is $(5.4 \times 10^6, 9.9 \times 10^6)$.
Although this interval is much narrower than the standard
interval and at first glance its limits seem sensible, the
interval fails to include the true Y_+, 13.8×10^6!
Further, from Stem-and-Leaf 1, even the 99% interval based
on all 100 simulated values of Y_+, $(5.2 \times 10^6$,
$11.8 \times 10^6)$, excludes the true value of Y_+ by a large
amount as well as the estimate based on the sample mean,
$N \times \bar{y} = 15.8 \times 10^6$. Of particular importance, this failure
to include the population total occurs with a sample that
from Table 1 appears quite representative of the population. For this sample, the inference for Y_+ based on the
log normal specification is, at least for the practical man
with hindsight, worse than the simple standard inference
for Y_+.

A re-examination of Figure 2 suggests one possible
reason for our excluding the right answer when using the log
normal specification: although $\log(Y_i)$ is substantially
more normal than Y_i, the 100 values of $\log(Y_i)$ are

STEM-and-LEAF 1:   The posterior predictive distribution
of Y_+ in Sample 1 based on a normal model for $\log(Y_i)$;
100 simulated values in units of 10^6.

```
 5.   124455778888999
 6.   00000111222222233344556666777777788888999
 7.   000011123344445557778899
 8.   0112233555667
 9.   01234
10.   3
11.   38
```

STEM-and-LEAF 2:   The posterior predictive distribution
of Y_+ in Sample 1 based on a normal model for $Y_i^{-1/8}$;
100 simulated values in units of 10^6.

```
 5.   56899
 6.   023335566668999
 7.   0011236789
 8.   022334555888889
 9.   33455668999
10.   0112344466
11.   002356
12.   347
13.   457899
14.   26
15.   77
16.   23
17.   223
18.   0
```

High values   21.3, 21.1, 26.0, 27.1, 30.1, 31.8, 32.5, 53.5

still not really normally distributed. In particular, a
straight line in the log transformation probability plot
goes well below the largest observed value. As a consequence, values like the largest observed value will be
generated less often by the log normal model than once in
one-hundred, with the result that the total as estimated
under the log normal specification will be relatively
small. Perhaps another transformation that produced
straighter probability plots would have led to better
results.

Before considering other transformations, we note that
the example illustrates the first point mentioned in the
Section 2. Although the log normal seems to fit the data
fairly well in a global sense as judged by the probability
plot, the inference for the total seriously underestimates
the actual total. Such behavior is not desirable when
trying to estimate total amounts of pollutant, radiation,
medical expenses or oil reserves, all examples which at
times are handled by log normal specifications.

4. SAMPLE 1 -- EXTENDED ANALYSES

Box and Cox (1964) suggest that the following family of
power transformations indexed by λ can be useful in
Bayesian and likelihood data analyses:

$$Z_i = \begin{cases} Y_i^\lambda & \text{if } \lambda \neq 0 \\ \log(Y_i) & \text{if } \lambda = 0 \end{cases}$$

where the Z_i are then assumed to be i.i.d. $N(\mu, \sigma^2)$. With
a particular choice of noninformative prior distribution on
(λ, μ, σ^2), the posterior distribution of λ is
proportional to

$$\text{Var}(Z_*)^{-(n-1)/2}$$

where

$$
Z_{*i} = \begin{cases} (Y_i^\lambda - 1)/(\dot{y}^{\lambda-1}) & \text{if} \quad \lambda \neq 0 \\[2ex] \dot{y} \log(Y_i) & \text{if} \quad \lambda = 0 \end{cases}
$$

$\dot{y}$ = geometric mean Y_i ,

and $\text{Var}(Z_*) = \sum (Z_{*i} - \overline{Z}_*)/(n - 1)$.

Table 2 presents values of $\text{Var}(Z_*)$ for twelve values of λ. Quite clearly, $\lambda = -1/8$ or even $\lambda = -1/4$ gives a substantially better fit to normality in Sample 1 than $\lambda = 0$. Figure 3 gives the normal probability plot of the sample values, $Y_i^{-1/8}$. Although it is not a straight line, the plot does seem somewhat straighter than the corresponding one for $\log(Y_i)$.

The same technique used to simulate the posterior distribution of Y_+ when $Z_i = \log(Y_i)$ was assumed normal, was used to simulate the posterior distribution of Y_+ when $Z_i = Y_i^{-1/8}$ was assumed normal: simply let $Z_i = Y_i^{-1/8}$ instead of $\log(Y_i)$ and let $Y_i = Z_i^{-8}$ instead of $\exp(Z_i)$. One problem that has to be addressed in principle, and possibly in practice, is that negative values of Z_i are possible because Z_i is assumed to be normally distributed, and negative Z_i values do not map properly into Y_i values. Formally, we will assume that Z_i is distributed as a truncated normal; thus, if a negative value of Z_i is generated, we will draw a new Z_i value; the Appendix provides details.

Based on the 100 simulated values displayed in Stem-and-Leaf 2, the posterior median of Y_+ is 9.6×10^6, and the 95% interval based on the 3rd and 97th values is $(5.8 \times 10^6, 31.8 \times 10^6)$. Note that the interval includes the true value, that the upper limit is similar to the upper limit of the standard interval but that the lower limit is closer to the true value.

Perhaps we have learned how to successfully apply likelihood/Bayesian methods with such data - use the Box-Cox family of power transformations as the basic model with

TABLE 2: Fit of Power Family:

$$\text{Var}(Z_*) \times 10^{-7}$$

Power	Sample 1	Sample 2
1	2022.57	5226.94
1/2	14.06	30.84
1/4	2.58	4.55
1/8	1.59	2.43
1/16	1.37	1.95
1/32	1.29	1.78
log	1.23	1.65
-1/32	1.18	1.55
-1/16	1.15	1.48
-1/8	1.11	1.37
-1/4	1.13	1.32
-1/2	1.47	1.64

$$Z_* = \begin{cases} (y^\lambda - 1)/(\lambda \dot{y}^{\lambda-1}) & \lambda \neq 0 \\ \dot{y} \log(y) & \lambda = 0 \end{cases} \quad \text{where } \dot{y} = \text{geometric mean }(y).$$

With noninformative prior, posterior proportional to $\text{Var}(Z_*)^{-(n-1)/2}$

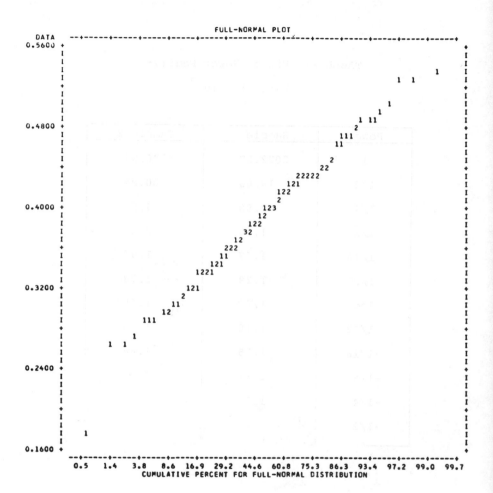

Figure 3: Normal Plot, $Y_i^{-1/8}$, Sample 1.

simulation techniques as the computational tool. But we did
not conduct a very rigorous test of this conjecture. We
started with the log transformation and obtained an inference that looked respectable but excluded the true value, a
fact never known in practice; we then enlarged the family of
transformations and found the best fitting transformation.
This extended procedure seemed to work in the sense that the
resultant 95% interval was plausible and covered the true
value. To check on this extended procedure, we will try it
on a second random sample of 100. This second sample was
the only other one selected.

5. SAMPLE 2

The second sample of 100 cities and towns is summarized
in Table 1. The standard inference for the population total
from this sample is that $(-3.4 \times 10^6 \times 65.3 \times 10^6)$ is a
95% interval. Substituting the sample total for the lower
limit gives $(3.9 \times 10^6, 65.3 \times 10^6)$, a large interval
which includes the true value.

The Sample 2 data were first modelled as log normal,
and 100 values were drawn from the posterior distribution of
the total. The resultant posterior median is 10.6×10^6,
and the 95% interval based on the third and 97th simulated
values is $(8.2 \times 10^6, 19.6 \times 10^6)$; the 99% interval
based on the lowest and highest simulated values is
$(8.1 \times 10^6, 25.3 \times 10^6)$. The log normal inference is quite
tight and covers the true value, although not the estimate
based on the sample mean, $N\bar{y} \doteq 31 \times 10^6$. If we had drawn
Sample 2 first, we might have concluded that the log normal
model for this population was perfectly satisfactory. But
based upon our experience with Sample 1, we should not trust
the log normal interval and instead should consider the
power family. Figure 4 shows that the log normal does not
provide an entirely satisfactory fit to Sample 2 just as it
did not to Sample 1. In fact, judging from the normal
plots, the log normal fits more poorly in Sample 2 than in
Sample 1 even though with pragmatic hindsight, the 95%

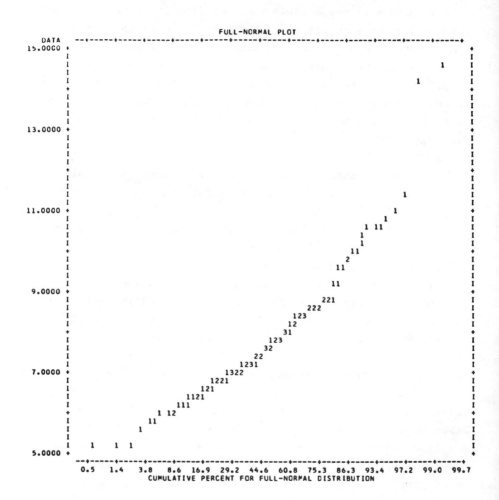

Figure 4:   Normal Plot,   Log(Y_i),   Sample 2.

interval for Y_+ in Sample 2 is more satisfactory than the
95% interval for Y_+ in Sample 1.

Values of $Var(Z_*)$ for sample 2 are given in Table 2.
As with sample 1, the log is not the best transformation;
now, $\lambda = -1/4$ is best, slightly better than $\lambda = -1/8$.
Figures 5 and 6 show the normal probability plots for
$Z_i = Y_i^{-1/4}$ and $Z_i = Y_i^{-1/8}$ respectively for sample 2;
both transformations appear better than the log
transformation.

Even though the sampled values of $Y^{-1/4}$ appear to be
rather normal, the inferences for the population total
resulting from assuming that $Z_i = Y_i^{-1/4}$ follow a truncated
normal distributed are, with pragmatic hindsight, atrocious:
all 100 generated values of Y_+ are larger than the true
value of Y_+ and most of them are much larger. In fact,
the resulting 100 draws from the posterior distribution
for Y_+ is so long-tailed that it is not well-summarized
by a stem-and-leaf display: the minimum value generated is
14.1×10^6, the third lowest is 18×10^6, the median
is 57×10^7, the 97th value is 14×10^{15} and the largest
value generated is 12×10^{17}! The best value for λ
yields entirely unsatisfactory inferences for Y_+: the 99%
interval is extremely large and excludes the correct answer.

The inferences that result from using $\lambda = -1/8$ are,
from a practical point of view, substantially better
although still not very satisfying: the posterior median is
15.7×10^6 and the 95% interval based on the third and
97th values is $(8 \times 10^6, 200 \times 10^6)$. Although in Sample 2
both $Y^{-1/8}$ and $Y^{-1/4}$ are better transformations to
normality than $\log(Y_i)$, at least judging by likelihood
criteria and probability plots, the inferences for Y_+
under these models are far worse than the inferences for
Y_+ under the log normal model, at least to the practical
man who wants a tight interval that covers the true value.
These results illustrate the second point in Section 2.

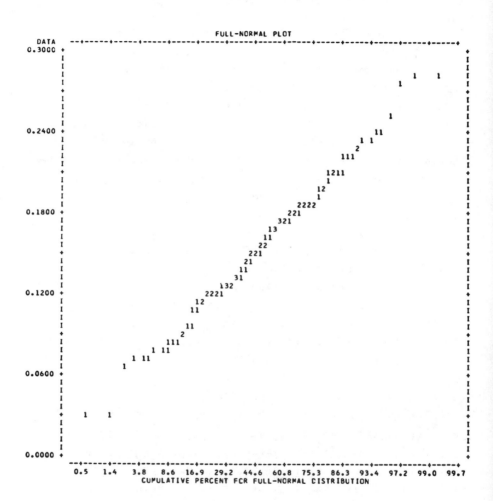

Figure 5: Normal Plot, $Y_i^{-1/4}$, Sample 2.

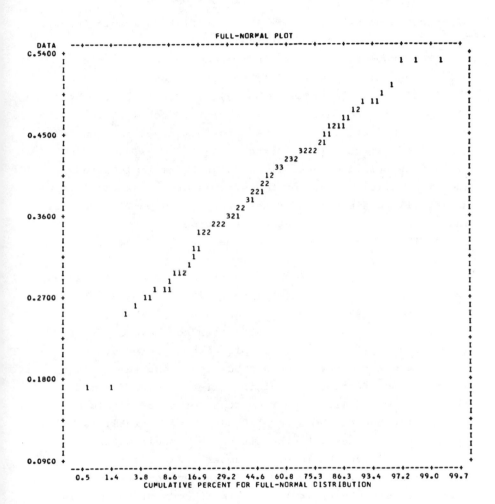

Figure 6: Normal Plot, $Y_i^{-1/8}$, Sample 2.

6. NEED TO SPECIFY CRITICAL PRIOR INFORMATION

What's going on? How can the inferences for the population total in Sample 2 be so much less realistic with better fitting models (e.g., with $Y_i^{-1/8}$ and $Y_i^{-1/4}$ distributed normally) than with worse fitting models (e.g., with $\log(Y_i)$ distributed normally)?

The problem with these inferences in this example is not an inability of the models to fit the data. A larger family of transformations to normality that could further straighten the normal probability plot is not what is needed. In fact, all monotone transformations that map the ith order statistic $Y_{(i)}$ into $\mu + \sigma \Phi^{-1}\left(\frac{i}{n+1}\right)$ for any μ and σ yield essentially straight normal plots and identical likelihoods, yet these transformations can lead to drastically different inferences for Y_+ depending on their shape for values of Y between the order statistics and especially for values of Y greater than the largest order statistic, $Y_{(n)}$. There exists an infinity of such transformations and none can be contradicted by or selected by probability plots or likelihood criteria alone. The problem is that the question we are asking, "What is the total, Y_+, in the population?", does not have a stable answer from a simple random sample without information external to the observed data about the right tail of the distribution of sizes of municipalities. As we fit models like the power family, the right tail of these models, (especially beyond the upper 1/2 percentage point), is being wagged uncontrollably by the fit of the model to the body of the data (between the lower and upper 1/2 percentage points); behavior of the models in the extreme tails is not being addressed by the relative likelihoods of the models (or by the corresponding probability plots) because there are no data in the extreme tails. Yet the inference for Y_+ is critically dependent upon tail behavior beyond the percentile corresponding to the largest observed Y_i. In order to estimate the total, not only do we need a model that provides a reasonable fit to the observed data, we also

need a model that provides realistic extrapolations beyond
the region of the data. For such extrapolations, we must
rely on prior assumptions, such as specification of the
largest possible size of a municipality.

More explicitly, for our two samples, the three
parameters of the power family, λ, μ, σ^2, are basically
enough to provide a reasonable fit to the observed data;
$\lambda = -1/8$ in Sample 1 and $\lambda = -1/4$ in Sample 2 pretty
much generate straight probability plots. But in order to
obtain realistic inferences for the population of New York
State from both samples, we need to constrain the
distribution of large municipalities. Suppose that a priori
we know that no city has population greater than 5×10^6.
Then using the simulation techniques described in the
Appendix, we can draw values from the posterior distribution
of size of municipality truncated at 5×10^6. Stem-and-
Leafs 3 and 4 display the resultant posterior distributions
of Y_+ from Samples 1 and 2 using the best fitting power
for each ($\lambda = -1/8$ and $\lambda = -1/4$ respectively) and
truncating the size of municipality at 5×10^6. Although
this method of providing prior information may seem somewhat
clumsy, these Stem-and-Leaf displays yield quite reasonable
inferences for the total population size; in both samples,
the inferences for Y_+ are tighter than with the
untruncated models and in Sample 2, the inference is
realistic. In both samples, the 95% intervals cover the
true value: the interval in Sample 1 is $(6 \times 10^6,$
$20 \times 10^6)$ and the interval in Sample 2 is
$(10 \times 10^6, 34 \times 10^6)$.

The point is simple, and was stated in Section 2: if
we ask a question and wish good statistical answers from the
data at hand, we must in general provide models that (a) are
flexible enough to let the data fit features it can (e.g.,
the power family of transformations to normality is nearly
flexible enough to generate straight probability plots for
our data), and (b) impose prior constraints on critical
features of the underlying distribution that the data cannot

STEM-and-LEAF 3:   The posterior predictive distribution
of Y_+ in Sample 1 based on a truncated normal model for
$Y_i^{-1/8}$, $Y_i < 5 \times 10^6$; 100 simulated values in units of 10^6.

```
 5.   56899
 6.   0233355666678999
 7.   0011234567789
 8.   0022233445556888889
 9.   33455668999
10.   011123444466
11.   0012356
12.   34
13.   24789
14.
15.   78
16.   24
17.   122
```

High values: 20.0, 24.8, 26.0

STEM-and-LEAF 4:   The posterior predictive distribution
of Y_+ in Sample 2 based on a truncated normal model for
$Y_i^{-1/4}$, $Y_i < 5 \times 10^6$; 100 simulated values in units of 10^7.

```
0.   88
1.   011
1.   2222223333333
1.   4444555555
1.   6666777777
1.   88888999999
2.   0000011111
2.   223333333333
2.   444444444445
2.   666777
2.   8
3.   000011
3.   3
3.   45
3.
3.   8
```

address (e.g., restrict all municipality sizes to be less
than 5×10^6).

7. GOOD FITS AND SPECIFIED EXTREME VALUES ARE NOT ENOUGH
WITH SUCH DATA

The results in the previous section might be seen as
suggesting that in order to estimate the population total
from such data, it is sufficient to (a) apply a
transformation that produces a basically straight
probability plot and (b) specify the smallest and largest
possible values. This conclusion would be incorrect,
however, because inferences for Y_+ are still sensitive to
the particular shape of the implied distribution of Y
between the order statistics, and once again the data cannot
distinguish between the alternatives. Two rather ad hoc
inferential techniques will be used to demonstrate this
fact.

The first method applies an ad hoc transformation to
the Y_i that produces an essentially straight normal
probability plot. The method is similar to the use of power
transformations in that a transformation is found that
straightens the probability plot and then the transformation
is regarded as known; it differs from the family of power
transformations in that it fits, in some sense, $n - 1$
parameters rather than 1. The procedure for our data is as
follows: map Y_i into $\phi^{-1}(\frac{i}{101})$ $i = 1,...,100$; map
$Y_{max} = 5 \times 10^6$ into 4 and $Y_{min} = 1$ into -4; linearly
interpolate between these points and truncate at Y_{min}
and Y_{max}. This procedure produces essentially straight
probability plots and truncates at realistic values, yet the
resulting inferences for Y_+ are quite different from the
inferences for Y_+ based on the truncated $Y_i^{-1/8}$
transformation in Sample 1 or the truncated $Y_i^{-1/4}$
transformation in Sample 2, primarily because of the shape
of the transformation between the large order statistics and
between $Y_{(n)}$ and Y_{max}: the resultant 95% interval for
Y_+ from Sample 1 is $(10 \times 10^6, 57 \times 10^6)$ and from Sample
2 is $(19 \times 10^6, 108 \times 10^6)$.

Relative to this ad hoc transformation, the power transformations smoothed the tails of the implied distributions for Y, and, in Sample 2, thereby discounted to some extent the fact that the two largest order statistics were similar and substantially larger than the other 98 values.

The second rather ad hoc method of inference for Y_+ used here is the Bayesian Bootstrap (Rubin, 1981), which places an improper Dirichlet prior distribution over all possible values, with the result that unobserved values have zero posterior probability and observed values are equally likely. Although not a transformation to normality, it implies a population distribution that perfectly reflects the sample distribution and so is like a transformation to normality with a straight normal probability plot. Note, however, the extreme form of the implied distribution of Y between the order statistics: all mass is concentrated at the order statistics, a vastly different assumption from the previous one which spread out the probability from $Y_{(i)}$ to $Y_{(i+1)}$ according to a linear interpolation rule. Applying the Bayesian Bootstrap to Sample 1 and Sample 2 yields simulated 95% intervals equal to $(4 \times 10^6, 49 \times 10^6)$ and $(7 \times 10^6, 81 \times 10^6)$ respectively. These intervals are respectable, although not particularly sharp, even though the prior specification on which they are based is absurd in that it leads to all posterior mass concentrated at the observed values.

The intervals based on the truncated power transformation, the ad hoc linear interpolation transformation, and the Bayesian Bootstrap are not extremely similar to each other. Consequently, having a model that provides a perfect fit to our data is not enough to draw robust inferences for the population total, even if supplemented with prior specification of extreme values. The inferences are still somewhat sensitive to the shape of the population distribution between the large order statistics implied by the specified transformation.

8. ROBUST QUESTIONS AND SAMPLES OBVIATE THE NEED FOR STRONG
   PRIOR INFORMATION

Of course, simulation techniques are not needed to
estimate totals routinely in practice. Good survey
practitioners know that a simple random sample is not a good
survey design for estimating the total in a highly skewed
population. If stratification variables were available
(e.g., that categorized municipalities into villages, towns,
cities, and boroughs of New York City), in order to estimate
the population total from a sample of 100, oversampling the
large municipalities would be highly desirable (e.g., sample
all five boroughs of New York City, many cities, several
towns, and a few villages).

It should not be overlooked, however, that the simple
random samples we drew, although not ideal for estimating
the population total, are quite satisfactory for answering
many questions without imposing strong prior restrictions.
Such questions are robust for our simple random samples in
the sense that their answers are relatively stable over a
broad range of plausible models. Robustness in this sense
is a joint property of questions, data, and models that are
not contradicted by observed data.

Table 3 illustrates the relative robustness of
inference for interior percentiles from our data. Even with
extreme interior percentiles and poorer fitting transformations, the resulting inferences are usually realistic.
Better models tend to give better answers, but for questions
such as these that are robust for the data at hand, the
effect is rather weak: For these questions, prior
constraints are not extremely critical and even relatively
inflexible models can provide satisfactory answers. Of
course, other robust questions would have been the value of
the population mid-mean or some other population analogue of
a robust-statistic.

The critical issue being illustrated is that robustness
is not a property of data alone or questions alone, but
particular combinations of data, questions and families of

TABLE 3:   SIMULATED POSTERIOR DISTRIBUTIONS FOR PERCENTILES
Based on 100 Draws and Various Transformations to Normality

Population Percentile		Sample 1				Sample 2			
		Log	-1/8	-1/4	$-1/8^T$	Log	-1/8	-1/4	$-1/4^T$
$5\text{th} \div 10^2$ 3.4	Low	1.1	1.9	2.3	1.8	1.0	1.1	1.6	1.6
	3rd	1.2	2.0	2.4	1.9	1.1	1.5	2.0	2.0
	Med^n	2.2	2.9	3.2	2.9	1.7	2.5	3.0	3.0
	97th	3.1	3.5	3.8	3.5	2.6	3.4	3.7	3.7
	High	3.1	3.8	4.0	3.8	2.7	3.6	3.9	3.9
$25\text{th} \div 10^2$ 8.0	Low	5.9	6.1	6.1	6.1	5.2	4.8	5.1	5.1
	3rd	6.4	6.7	6.4	6.6	5.6	5.5	5.6	5.5
	Med^n	8.8	8.8	8.5	8.8	7.8	8.0	7.8	7.8
	97th	10.9	10.7	10.2	10.7	10.8	10.6	10.0	9.9
	High	11.0	11.1	11.1	11.5	11.8	11.0	10.3	10.2
$\text{Med}^n \div 10^3$ 1.7	Low	1.7	1.4	1.3	1.4	1.5	1.3	1.2	1.2
	3rd	1.8	1.6	1.5	1.6	1.7	1.4	1.3	1.2
	Med^n	2.3	2.1	1.9	2.1	2.3	2.1	1.8	1.8
	97th	3.0	2.7	2.4	2.7	3.6	2.8	2.4	2.4
	High	3.6	2.8	2.5	2.8	4.0	3.0	2.6	2.6
$75\text{th} \div 10^3$ 5.1	Low	4.4	4.3	3.9	4.3	4.9	4.0	3.5	3.4
	3rd	4.7	4.3	3.9	4.3	4.9	4.4	3.7	3.6
	Med^n	6.2	5.7	5.2	5.7	7.1	6.0	5.3	5.2
	97th	9.1	7.9	7.3	7.9	12.3	8.6	7.8	7.2
	High	9.9	8.9	8.2	8.8	14.6	9.4	8.2	8.0
$95\text{th} \div 10^3$ 30.3	Low	19	19	22	19	23	22	23	23
	3rd	20	20	22	20	26	24	27	26
	Med^n	26	29	39	29	38	40	48	45
	97th	45	64	112	64	76	61	113	86
	High	47	77	133	74	128	75	127	93

T = truncated at 5×10^6

models. In many problems, statisticians may be able to
define the questions being studied so as to have robust
answers. We, in fact, did this by summarizing simulated
posterior distributions by percentiles rather than
moments. Often, however, the practical, important question
is inescapably nonrobust. To repeat the central theme of
this article: statisticians have an obligation to provide
the kinds of answers clients will assume are being provided
along with appraisals of the sensitivity of the inferences
to assumptions unassailable by the data; we must face the
fact that, in general, inferences rely on assumptions that
the data at hand cannot address.

APPENDIX

1. Notation

Y_i i = 1,...,N are the N values of Y in the
population.

A priori $\ell < Y_i < u$; e.g., $(0,\infty)$, $(2, 5 \times 10^6)$.

Y_i i = 1,...,n n < N are the known values of
Y in the sample.

$f(\cdot)$ is the normalizing transformation, $Z_i = f(Y_i)$.

$$Z_i = f(Y_i), \quad L < Z_i < U, \quad L = f(\ell), \quad U = f(u)$$

$$\bar{Z} = \sum_1^n Z_i/n$$

$$s_z^2 = \sum_1^n (Z_i - \bar{Z})/(n - 1)$$

2. Distributions

   Given parameters (μ, σ),   we assume

$$
z_i \overset{i.i.d}{\sim} \begin{cases} \frac{1}{\sqrt{2\pi}\sigma} \exp\left[-\frac{1}{2}\left(\frac{z_i - \mu}{\sigma}\right)^2\right]/k_L^U(\mu,\sigma) & \text{if } L < z_i < U \\[2ex] 0 & \text{otherwise} \end{cases}
$$

where $k_L^U(\mu,\sigma) = \int_L^U \frac{1}{\sqrt{2\pi}\sigma} \exp\left[-\frac{1}{2}\left(\frac{t-\mu}{\sigma}\right)^2\right]dt$.

With prior distribution $p(\mu,\sigma)$ for (μ,σ),   the posterior
distribution of (μ,σ) is proportional to

$$
\begin{cases} p(\mu,\sigma)k_L^U(\mu,\sigma)^{-n}\sigma^{-n}\exp\left[-\frac{1}{2}\sum_1^n \left(\frac{z_i - \mu}{\sigma}\right)^2\right] & \text{if all } L < z_i < U \\[2ex] 0 & \text{otherwise .} \end{cases}
$$

Assuming

$$
p(\mu,\sigma) \propto k_L^U(\mu,\sigma)^n/\sigma , \tag{A1}
$$

the posterior distribution of (μ,σ) is the same as with
the usual "noninformative" prior distribution for (μ,σ)
when $L = -\infty$ and $U = +\infty$.

   For most values of L, U, μ and σ in the simulation
presented here, $k_L^U(\mu,\sigma)^n \doteq 1$,   so that usually the choice
of the convenience prior distribution (A1) is not
substantially different from the more standard choice
proportional to σ^{-1}.

3. Simulation Loop

   Each pass through the following three steps produces
one draw from the posterior predictive distribution of
population quantity.

Step 1 - Draw μ, σ from their posterior distribution

$$
\sigma_*^2 = n s_z^2/\chi_{n-1}^2, \quad \chi_{n-1}^2 \text{ a } \chi^2 \text{ variate on } n - 1 \text{ df.}
$$

$$
\mu_* = \bar{z} + \sigma_* \times N(0,1)/\sqrt{n}, \quad N(0,1) \text{ a standard normal.}
$$

Step 2 – Draw unobserved Y_i from posterior predictive distribution given $\mu = \mu_*$ and $\sigma = \sigma_*$.

For $i = n + 1, \ldots, N$:

$$Z_i = \mu_* + \sigma_* \times N(0,1)$$

If $L < Z_i < U$, $Y_i = f^{-1}(Z_i)$; otherwise, redraw $N(0,1)$.

Step 3 – Calculate population quantity

$$\text{E.g. population total} = \sum_1^N Y_i$$

$$\text{population median} = \text{median } \{Y_1, \ldots, Y_N\}.$$

REFERENCES

1. Box, G. E. P. and D. R. Cox (1964). An Analysis of Transformations. J. Roy. Statist. Soc. B, 26, 211.

2. Cochran, W. G. (1963). Sampling Techniques 2nd ed., New York, Wiley.

3. Efron, B. (1980). Bootstrap Methods: Another Look at the Jackknife. Ann. Statist., 7, 1-26.

4. Ericson, W. A. (1969). Subjective Bayesian Models in Sampling Finite Populations. J. Roy. Statist. Soc. B, 31, 195-224.

5. Miller, R. G. (1974). The Jackknife – a review. Biometrika 61, 1-15.

6. Rubin, D. B. (1978). The Phenomenological Bayesian Perspective in Sample Surveys from Finite Populations: Foundations. Imputation and Editing of Faulty or Missing Survey Data. U. S. Department of Commerce, pp. 10-18.

7. Rubin, D. B. (1981). The Bayesian Bootstrap. Ann. Statist., 9, 130-134.

ACKNOWLEDGEMENTS

I wish to thank P. R. Rosenbaum for helpful comments on earlier drafts of this paper and D. T. Thayer for programming support. The work was partially supported by the Educational Testing Service and the Mathematics Research Center.

Mathematics Research Center
University of Wisconsin-Madison
Madison, WI 53706

Estimation of Variance of the Ratio Estimator: An Empirical Study

C. F. Jeff Wu and L. Y. Deng

1. INTRODUCTION

This paper concerns the estimation of variance of the ratio estimator under simple random sampling. While the setting is simple, we hope this study will eventually lead to better understanding of the important problem of variance estimation in complex surveys. In fact even in this simple setting, the problem of choosing "good" variance estimators is unsettled. More than a dozen estimators, proposed in a span of some thirty years, are listed in Rao (1969) and Royall and Cumberland (1981). The majority of estimators are design-based, i.e., their justification and choice are based on the performance according to the probability mechanism that generates the sample. A few others, proposed by Royall and his collaborators, are model-based. According to this approach, the inference should be made conditional on the observed sample and a hypothetical superpopulation model. The sampling design becomes irrelevant. Other estimators, e.g. the jackknife, may not be justified exclusively by either approach.

Previous work on the comparison of variance estimators for ratio include, among others, Rao and Beegle (1967), Rao (1968, 1969), Rao and Rao (1971), Rao and Kuzik (1974), Royall and Eberhardt (1975), Royall and Cumberland (1978, 1981), Krewski and Chakrabarty (1981), and Wu (1982). The theoretical comparison of various variance estimators is

made by assuming that the x and y populations satisfy
some linear regression models (superpopulations). Although
the results are sometimes exact, dependence on the
superpopulation parameters can be delicate. This prompted
Royall and Eberhardt (1975) to study the model-robustness of
some variance estimators. Their definition of robustness is
restricted to the bias behavior of the variance estimators
when the true parameters deviate from the assumed ones. Wu
(1982) gave the first model-free comparison of some variance
estimators by expanding the estimators and working on the
leading terms of the expansion. Such a comparison is large
sample in nature. On the empirical side, comparison is
conducted on either natural populations or artificial
populations simulated according to some superpopulation
models. The bias and/or mean square error of the variance
estimators are noted. Motivated by the prediction theory
approach, Royall and Cumberland (1981) took a different
approach by studying the conditional behavior of the
variance estimators as a function of the x-sample mean $\bar{x}$.
They showed that some variance estimators can behave
drastically different over a range of the $\bar{x}$ values. They
further argue that the conditional (on $\bar{x}$) mean of a
variance estimator should closely follow the conditional (on
$\bar{x}$) mean square error of the ratio estimator. We will come
back to this point in §5 and §6.

The work to be presented is empirical and is in part
inspired by the very stimulating paper of Royall and
Cumberland (1981). Their approach to empirical study can be
further improved in three respects. They did not consider
estimator v_2 defined in §2, which is motivated by the
probability sampling theory and is popular in practice. In
any effort to criticize the more traditional sampling theory
approach, it seems fair to consider both v_0 and v_2. See
also J. N. K. Rao's discussion of Royall and Cumberland
(1978). It will be shown later that v_2 is better than
v_0 in several desirable respects. Their conclusion in
favor of the prediction theory approach could have been more
convincing had they included the stronger "rival" v_2 in
their study. To remedy this we have included several

additional estimators in our study. The six natural
populations they chose look artificial in that they are all
well fitted by model (2), i.e. straight lines through the
origin with increasing residuals. We consider nine
populations, six identical to theirs and three incorporating
violations of three key assumptions of the linear regression
model (2) that typically underlines the use of ratio
estimator. Detail is in §4. Besides studying the
conditional behavior of the variance estimators in
"tracking" the conditional MSE of the ratio estimator, an
innovation due to Royall and Cumberland, we also study the
bias and MSE of the variance estimators as estimators of the
unconditional MSE of the ratio estimator and, more
importantly, the actual coverage probabilities of the
associated interval estimates of the y-population mean $\bar{Y}$
as compared with the nominal ones.

2. RATIO ESTIMATOR AND ITS VARIANCE

     Suppose that a population consists of N distinct
units with values (y_i, x_i), where $x_i > 0$ for $1 < i < N$.
A simple random sample of size n is taken without
replacement from the population. Denote the sample and
population means of y_i and x_i by $\bar{y}, \bar{x}$ and $\bar{Y}, \bar{X}$
respectively. The ratio estimator

$$\hat{\bar{y}}_R = \bar{y}\,\frac{\bar{X}}{\bar{x}} \tag{1}$$

is a popular estimator of $\bar{Y}$. It is simple to use in
practice. It combines efficiently the covariate information
in x_i when y_i and x_i are roughly positively
correlated. It is the best linear unbiased predictor of $\bar{Y}$
under the following superpopulation model (Brewer, 1963;
Royall, 1970)

$$y_i = \beta x_i + \varepsilon_i , \tag{2}$$

where ε_i are independent with mean zero and variance
$\sigma^2 x_i$. The ratio estimator possesses other desirable
properties. For example, it is robust against extreme
values in the individual ratios y_i/x_i (Rao, 1978).
Traditionally the ratio estimator is favored over the

regression estimator mainly for computational ease in handling large data sets. Given the present capacity of computers this should be less of a concern. Fuller (1977) gave examples to show that ratio estimation can be much less efficient than regression estimation. We believe it is time that more attention should be given to regression estimation.

There is no closed form for MSE $\hat{\bar{y}}_R$ or Var $\hat{\bar{y}}_R$. Both can be approximated by the approximate variance (Cochran, 1977, p. 155)

$$V_{appr} = \frac{1-f}{n} \frac{1}{N-1} \sum_1^N \left(y_i - \frac{\bar{Y}}{\bar{X}} x_i\right)^2 , \qquad (3)$$

where $f = n/N$ is the sampling fraction. For large samples the approximation is adequate. But for small sample size $(n < 12)$ V_{appr} can seriously underestimate MSE (Rao, 1968 or Cochran, 1977, p. 164). The most standard estimator of V_{appr} is its sample analogue

$$v_0 = \frac{1-f}{n} \frac{1}{n-1} \sum_1^n \left(y_i - \frac{\bar{y}}{\bar{x}} x_i\right)^2 . \qquad (4)$$

Some textbooks mention (but not endorse) v_2 as an alternative to v_0,

$$v_2 = \frac{1-f}{n} \left(\frac{\bar{x}}{\bar{x}}\right)^2 \frac{1}{n-1} \sum_1^n \left(y_i - \frac{\bar{y}}{\bar{x}} x_i\right)^2 . \qquad (5)$$

The original motivation for $v_2' = v_2/\bar{x}^2$ as a variance estimator of the ratio

$$R = \bar{Y}/\bar{X}$$

is the unavailability of $\bar{X}$. Both v_0 and v_2 are easy to compute.

3. VARIANCE ESTIMATORS UNDER STUDY

Let $e_i = y_i - Rx_i$ be the residual from the straight line connecting $(\bar{X},\bar{Y})$ and the origin, $\tilde{e}_i = y_i - rx_i$, $r = \bar{y}/\bar{x}$, be its sample analogue. Apart from a constant, V_{appr} is the population mean of the residual square e_i^2. Estimation of V_{appr} can be viewed as the more typical problem of estimating the population mean of a new characteristic e_i^2. By taking $e_i \approx \tilde{e}_i$, v_0 can be viewed as the sample mean of e_i^2 and should be less efficient than the ratio-type estimators v_2 or

$$v_1 = \frac{\bar{X}}{x} v_0 = \frac{1-f}{n} \frac{\bar{X}}{x} \frac{1}{n-1} \sum_1^n \tilde{e}_i^2 \tag{6}$$

when x_i and e_i^2 are positively correlated. A general class of estimators

$$v_g = \left(\frac{\bar{X}}{x}\right)^g v_0 \tag{7}$$

was proposed in Wu (1982). He proved that the leading term of $MSE(v_g)$ is minimized by

$$g_{opt} = \frac{S_{xz}\bar{X}}{S_x^2 \bar{z}}$$

$\qquad\qquad$ = population regression coefficient of $\qquad$ (8)

$$\frac{z_i}{\bar{z}} \quad \text{over} \quad \frac{x_i}{\bar{X}} \, ,$$

where

$$z_i = e_i^2 - 2 e_i \sum_1^N x_i e_i / \sum_1^N x_i \, ,$$

S_x^2 and S_{xz} are the population x-variance and (x,z)- covariance respectively. The second term of z_i accounts for the possible nonzero intercept in the population when fitted by a straight line according to (2). A (large-sample) model-free comparison of v_2 and v_0 readily obtains. When and only when $g_{opt} > 1$, v_2 is better than v_0. It may be easier to remember and to interpret the following approximation to g_{opt} (by ignoring the second term of z_i)

$\qquad$ g' = population regression coefficient of

$$\frac{e_i^2}{N^{-1}\sum_1^N e_i^2} \quad \text{over} \quad \frac{x_i}{\bar{X}} \, . \tag{9}$$

By taking a sample analogue to g_{opt}

$\hat{g}_{opt}$ = sample regression coefficient of

$$\frac{\tilde{z}_i}{\tilde{z}} \quad \text{over} \quad \frac{x_i}{\bar{x}} \, , \tag{10}$$

$$\tilde{z}_i = \hat{e}_i^2 - 2 \hat{e}_i \sum_1^n x_i \hat{e}_i / \sum_1^n x_i \, , \quad \tilde{z} = n^{-1} \sum_1^n \tilde{z}_i \, ,$$

we obtain an asymptotically optimal estimator $v_{\hat{g}_{opt}}$ within the class (7). Similarly we can take a sample analogue to g'

$\tilde{g}$ = sample regression coefficient of

$$\frac{\hat{e}_i^2}{n^{-1} \sum_1^n \hat{e}_i^2} \quad \text{over} \quad \frac{x_i}{\bar{x}} \tag{11}$$

and obtain another estimator $v_{\tilde{g}}$.

Instead of making a ratio adjustment to the sample mean of $\hat{e}_i^2$ as in v_1, Fuller (1981) suggested a regression adjustment to v_0. Denote his estimator by

$$v_{reg} = v_0 + \frac{1 - f}{n} \hat{b}_{e^2 x} (\bar{X} - \bar{x}) \tag{12}$$

where

$$\hat{b}_{e^2 x} = \text{sample regression coefficient of}$$

$$\hat{e}_i^2 \quad \text{over} \quad x_i \, .$$

By standard Taylor expansion, the leading term of $v_{\tilde{g}}$ is v_{reg} and their asymptotic behaviors should be close.

Another estimator of interest is the jackknife variance estimator

$$v_J = (1 - f) \bar{x}^2 \frac{n - 1}{n} \sum_1^n D_{(j)}^2 \, , \tag{13}$$

where $D_{(j)}$ is the difference between the ratio $(n\bar{y} - y_j)/(n\bar{x} - x_j)$ and the average of these n ratios. Royall and Cumberland (1981) and Krewski and Chakrabarty

(1981) studied the model-based and sampling properties of v_J. Note that the usual justification of jackknife is independent of a superpopulation model.

Royall and Eberhardt (1975) suggested

$$v_H = v_0 \frac{\bar{x}_c \bar{x}}{\bar{x}^2} \left(1 - \frac{c_x^2}{n}\right)^{-1} \tag{14}$$

when $\bar{x}_c$ = x-mean of non-sampled units, C_x = x-sample coefficient of variation. Later Royall and Cumberland (1978) suggested a closely related estimator

$$v_D = \frac{1 - f}{n} \frac{\bar{x}_c \bar{x}}{\bar{x}^2} \frac{1}{n} \sum_1^n \frac{\hat{e}_i^2}{1 - \frac{x_i}{n\bar{x}}} . \tag{15}$$

Both v_H and v_D are shown to be unbiased under model (2), approximately unbiased for more general variance patterns, and asymptotically equivalent to v_J.

Another variance estimator, which follows from standard least squares theory, is

$$v_L = \frac{1 - f}{n} \frac{\bar{x}_c \bar{x}}{\bar{x}^2} \frac{1}{n - 1} \sum_1^n \frac{\hat{e}_i^2}{x_i} .$$

It is unbiased under model (2) but can be seriously biased if $\text{var}(y_i) = \sigma^2 x_i$ in (2) is violated (Royall and Eberhardt, 1975). Their empirical behavior has been shown to be equally bad in Royall and Cumberland (1981). For these reasons v_L will not be considered in our study.

4. <u>POPULATIONS UNDER STUDY</u>

The preceding variance estimators are compared empirically on nine populations listed in Table 1. The first six are natural populations. The original data were generously provided to us by Professors W. G. Cumberland and R. M. Royall, to whom we wish to express our sincere thanks. For more detailed description of these populations, see their 1981 paper. The last three are transformations of population 1. Their description follows. The first six populations are plotted in Royall and Cumberland (1981, p. 69-70). Though being natural populations, they are all

TABLE 1. STUDY POPULATIONS

Population	Description	x	y
1	Counties in NC, SC, and GA with 1960 white female population <100,000	Adult white female population, 1960	Breast cancer mortality, 1950–69 (white females)
2	U.S. cities with 1960 population between 100,000 and 1,000,000	Population, 1960	Population, 1970
3	Counties in NC, SC, and GA with fewer than 100,000 households in 1960	Number of households, 1960	Population, excluding residents of group quarters, 1960
4	Counties in NC, SC, and GA with fewer than 100,000 households in 1960	Number of households, 1960	Population, excluding residents of group quarters, 1970
5	National sample of short-stay hospitals with fewer than 1,000 beds	Number of beds	Number of patients discharged
6	Corporations with 1974 gross sales between one-half billion and fifty billion dollars	Gross sales, 1974	Gross sales, 1975
7	Transformation of population 1 (see (17))		
8	Transformation of population 1 (see (18))		
9	Transformation of population 1 (see (19))		

For sources of populations 1 to 6, see Royall and Cumberland (1981, p. 68).

well described by straight lines through the origin using
weighted least squares. The squared residuals from the
fitted line increase roughly in proportion to x. More
refined models like $\alpha + \beta x$ (linear regression with
intercept) and $\alpha + \beta x + r x^2$ (quadratic regression) do not
differ significantly from the simpler linear-through-the
origin model βx except possibly for population 5. To
represent broader range of real populations, we construct
populations 7, 8 and 9 from population 1 to reflect the
violation of three key assumptions underlying the linear-
through-the origin model (2):   (i)   zero intercept, (ii)
 $var(y_i) \propto x_i$, (iii)   linearity of Ey_i in x_i. More
precisely, decompose the y_i value in population 1, denoted
old y_i, into

$$old \ y_i = Rx_i + (y_i - Rx_i)$$
$$= \hat{y}_i + e_i \ . \tag{16}$$

Define the new y_i value in population 7 as

$$new \ y_i = old \ y_i + \bar{Y} \ ; \tag{17}$$

for population 8,

$$new \ y_i = \hat{y}_i + kx_i e_i \ , \tag{18}$$

with $k = S_x^{-1}$ and all units except two have $y_i > 0$; for
population 9,

$$new \ y_i = c_0[c_1 - e^{\beta(x_i - \bar{X})}] + e_i \ , \tag{19}$$

where $\beta = S_x^{-1}$, $c_0 = S_y$, $c_1 = 0.1 + \exp[\beta(\max_{1 \leqslant i \leqslant N} x_i - \bar{X})]$
so that $y_i > 0$ for all i.

Populations 1, 7, 8 and 9 are shown below.

Some characterisics of the populations are given in
Table 2. Note that x and y are highly correlated
 (> 0.94) for populations 1-4, 6, 7. The x and y of
the transformed populations 8 and 9 are less correlated.
Another point to observe is that V_{appr} can be smaller or
larger than MSE for sample size 32. There is no systematic
pattern in the percent underestimate or overestimate (last

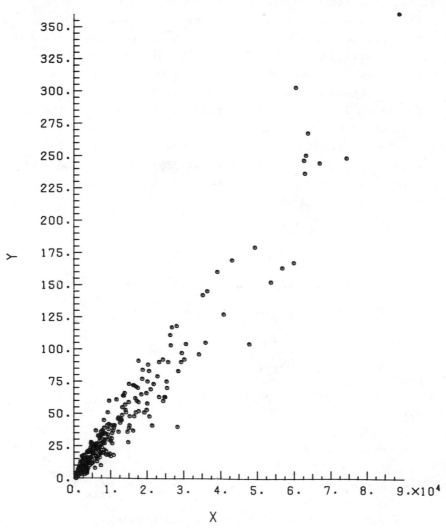

Figure 1.   Population (1)

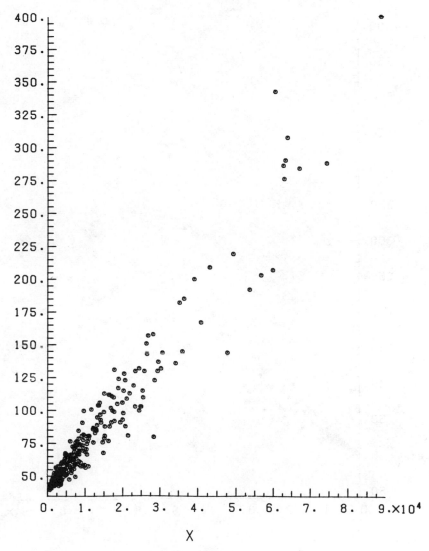

Figure 2.   Population (7)

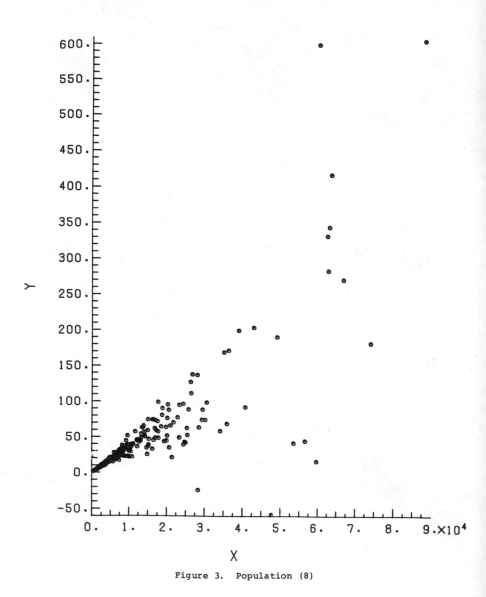

Figure 3. Population (8)

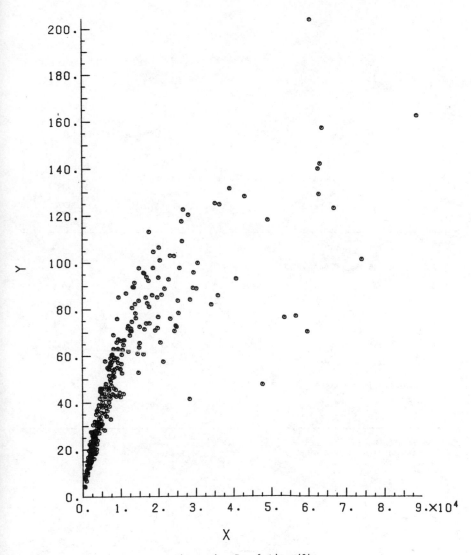

Figure 4. Population (9)

TABLE 2. SOME POPULATION CHARACTERISTICS

Population	N	$\rho(X,Y)$[a]	$MSE(\hat{\bar{y}}_R)$[b]	$V_{appr}(\hat{\bar{y}}_R)$	$\dfrac{100(V_{appr} - MSE)}{MSE}$
1	301	0.967	4.61	4.71	2.2
2	125	0.947	113.9×10^6	111.9×10^6	−1.8
3	304	0.998	33.6×10^4	32.4×10^4	−3.6
4	304	0.982	230.9×10^4	264.5×10^4	14.6
5	393	0.911	1941.0	1968.0	1.4
6	331	0.997	30.7×10^{14}	35.7×10^{14}	16.3
7	301	0.967	77.7	69.0	−11.2
8	301	0.805	44.3	47.6	7.4
9	301	0.824	44.6	43.0	−3.6

a. Correlation between x and y populations.
b. Based on 1000 simulated samples.

column of Table 2) in V_{appr} of MSE with only three of them
over ten percent. This is quite different from Rao (1968),
where he found that, for smaller sample sizes $n = 4, 6, 8,$
12, V_{appr} consistently underestimates MSE with average
percent underestimates ranging between 12% and 17%. The
discrepancy is explained in part by the difference in sample
sizes. More importantly, Rao's computation of V_{appr} and
MSE is apparently based on a particular superpopulation
model while ours is model free.

5. RESULTS

We draw 1000 simple random samples of size $n = 32$
from each population. For each sample we calculate the
ratio estimate $\hat{\bar{y}}_R$ and the variance estimates $v_0, v_1, v_2,$
$v_{\hat{g}_{opt}}, v_{\tilde{g}}, v_{reg}, v_J, v_H, v_D$ and $v_{g_{opt}}$. Note $v_{g_{opt}}$ is
not really an estimator since g_{opt} depends on the whole
population. We include it here to see how the asymptotic
results in Wu (1982) (or §3) predict the actual performance
for sample size 32. The $MSE(\bar{y}_R)$ in Table 2 is calculated
as $1000^{-1} \sum_1^{1000} (\hat{\bar{y}}_R - \bar{Y})^2$ over the 1000 simulated
samples. For each variance estimator v, its bias
bias(v) is calculated as $1000^{-1} \sum_1^{1000} v - MSE(\hat{\bar{y}}_R)$ over
the same 1000 samples, and its root mean-square error
$\sqrt{MSE}$ (v) as $(1000^{-1} \sum_1^{1000} (v - MSE(\hat{\bar{y}}_R))^2)^{1/2}$ over the same
1000 sample. Results are given in Table 3.

We first summarize the root mean square error behavior
of the ten estimators in Table 3 as follows.

(i) The asymptotically optimal estimator $v_{g_{opt}}$
(pretending g_{opt} is available) is the best or nearly the
best estimator in terms of minimizing MSE, as well predicted
by the asymptotic result of Wu (1982).

(ii) Among $v_0, v_1, v_2,$ the best performer is
consistently the one closer to g_{opt}. For example,
$g_{opt} = 1.59$ in population 1 is closer to 2 and thus v_2
has smaller $\sqrt{MSE}$ (2.20) than those (2.26 and 2.73) of
v_1 and v_0 respectively. This is again predicted by the
asymptotic result of Wu (1982, §2.2).

Table 3. Root mean-square error and bias* of variance estimators

variance estimator	\multicolumn{9}{c}{Population}								
	1	2	3	4	5	6	7	8	9
v_0	2.73 (-0.29)	54.3 (-0.7)	15.7 (-5.6)	262 (-13.5)	819 (-45)	24.9 (-0.6)	22.7 (-12.4)	42.1 (-7.3)	18.8 (-6.9)
v_1	2.26 (-0.43)	53.0 (-1.2)	13.9 (-6.4)	224 (-27.8)	731 (-71)	18.6 (-1.9)	24.1 (-10.6)	36.8 (-10.2)	17.3 (-7.6)
v_2	2.20 (-0.39)	56.6 (-0.2)	14.1 (-6.0)	214 (-32.1)	735 (-66)	17.1 (-1.0)	32.4 (-6.2)	35.0 (-11.9)	18.1 (-7.0)
$v_{\hat{g}_{opt}}$	2.26 (-0.44)	53.5 (-1.4)	14.7 (-4.3)	228 (-26.7)	755 (-65)	18.7 (-1.1)	23.6 (-10.4)	35.8 (-11.0)	18.0 (-5.5)
$v_{\tilde{g}}$	2.33 (-0.40)	53.5 (-1.2)	15.7 (-4.8)	243 (-25.4)	766 (-70)	18.6 (-1.4)	36.2 (-4.3)	37.7 (-10.5)	22.3 (-4.3)
v_{reg}	2.20 (-0.45)	52.3 (-1.3)	14.0 (-6.2)	209 (-35)	721 (-75)	17.4 (-2.3)	43.5 (-8.8)	35.2 (-11.5)	35.3 (-7.2)
v_H	2.27 (-0.25)	60.5 (3.4)	14.4 (-5.0)	220 (-25.7)	760 (-7)	17.8 (0.1)	34.6 (-3.3)	35.6 (-11.0)	18.5 (-5.7)
v_D	2.46 (-0.09)	59.5 (2.6)	15.8 (-2.9)	251 (-9.8)	782 (1)	21.3 (2.4)	38.9 (1.5)	39.0 (-7.6)	21.2 (-1.9)
v_J	2.76 (0.23)	58.3 (5.2)	18.6 (0.9)	303 (18.6)	829 (72)	28.0 (6.9)	45.4 (9.3)	45.0 (-2.2)	26.2 (4)
$v_{g_{opt}}$	2.17 (-0.43)	53.3 (-1.1)	13.7 (-6.4)	218 (-30.8)	722 (-72)	17.1 (-1.5)	22.6 (-12.4)	35.6 (-12.4)	17.2 (-7.6)
g_{opt}	1.59	1.18	1.24	2.46	1.60	1.72	0.06	2.80	1.05
unit	10^0	10^6	10^4	10^4	10^0	10^{14}	10^0	10^0	10^0

*Bias given inside the parenthesis

(iii) The estimators $v_{\hat{g}}$, $v_{\tilde{g}}$ and v_{reg} are asymptotically equivalent and are close to $v_{g_{opt}}$. They all give small $\sqrt{MSE}$. It is somewhat surprising that v_{reg} does as well as $v_{g_{opt}}$ and better than $v_{\hat{g}}$ and $v_{\tilde{g}}$ on populations 1-6 and 8. Reasons for the poor performance of v_{reg} on populations 7 and 9 are not known. The estimators $v_{\hat{g}}$, $v_{\tilde{g}}$ and v_H are more stable in that they perform reasonably well for all the populations.

(iv) The jackknife variance estimator v_J is the worst in terms of MSE. The instability of v_J was also reported in Rao and Rao (1971), Rao and Kuzik (1974), Krewski and Chakrabarty (1981). The performance of v_D is not good either.

The bias of each variance estimator is given inside the parenthesis in Table 3. The results are summarized as follows.

(i) The bias is usually a small proportion (say, < 30%) of the total $\sqrt{MSE}$ with a few exceptions for populations 3, 7, 8, 9.

(ii) The estimators v_0, v_1, v_2 are consistently downward biased for estimating the MSE. The estimators $v_{\hat{g}}$, $v_{\tilde{g}}$ and v_{reg}, being close to one of v_0, v_1 or v_2, are consistently downward biased. Another intriguing phenomenon: among v_0, v_1, v_2, those with smaller $\sqrt{MSE}$ tend to have bigger (in magnitude) bias.

(iii) The estimator v_J is almost always upward biased, while v_H and v_D exhibit no systematic pattern.

The downward biasedness of v_0 was noted in Rao (1968), Rao and Rao (1971). And the upward biasedness of v_J was noted in Rao and Rao (1971). Both are exact analytic results whose validity depends on some particular superpopulation models. Model-free (but asymptotic) results on the bias of v_0, v_1, v_2, v_J, v_H have been obtained by the first author. They will appear soon.

In estimating the population mean the purpose of variance estimation is rather for assessing the variability of the ratio estimator than for estimating the variance itself. A more interesting and relevant criterion is the

behavior of the associated confidence interval. For each
variance estimator v and each simulated sample, we
consider the t-statistic

$$t = \frac{\hat{\bar{y}}_R - \bar{Y}}{\sqrt{v}} ,\qquad(20)$$

and the $(1 - \alpha)$ confidence interval for estimating $\bar{Y}$

$$(\hat{\bar{y}}_R - t_{\alpha/2}(31)\ \sqrt{v},\ \hat{\bar{y}}_R + t_{\alpha/2}(31)\ \sqrt{v})\qquad(21)$$

where $t_{\alpha/2}(31)$ is the upper $\alpha/2$ point of the
t-distribution with d.f. = 31. The Monte-Carlo coverage
probability of the confidence interval (21), given in Table
4, is calculated as the percentage of the 1000 intervals
(21) that cover $\bar{Y}$. The bias, standard deviation and
coefficient of skewness of the associated t-statistic, given
in the last three columns of Table 4, are based on the 1000
t-values (20).

   We now summarize Table 4 in three parts: I. normality
of t-statistic, II. width of t-interval, III. reliability
of t-interval in terms of the closeness of its Monte Carlo
coverage probability to the nominal one.

   I. Except for populations 4 and 9, the bias is close
to zero, the s.d. close to one and the coefficient of
skewness close to zero. Typically the t-statistic
associated with the estimator v_0 is not normal, especially
with its large coefficient of skewness.

   II. Since the squared length of the t-interval is
proportional to the expected value of v, from E(v) = bias
of v + MSE, we can use the bias entry of Table 3 in
assessing the width of t-interval. Since v_J has positive
bias, the corresponding t-interval is wider. Similarly v_0,
v_1, v_2, $v_{\hat{g}}$, $v_{\tilde{g}}$, v_{reg} all have negative bias. Their
t-intervals are shorter. The intervals associated with
v_H and v_D are in between the two extremes.

   III. (i) Generally the coverage probability is lower
than the nominal level $1 - \alpha$. This may in part be
explained by the negative bias of v in most cases
(except v_J and some cases of v_H and v_D) The

Table 4. Coverage probabilities of the t-intervals $(\hat{\bar{y}}_R - c\sqrt{v},\ \hat{\bar{y}}_R + c\sqrt{v})$, $c = t_{\frac{\alpha}{2}}(31)$, and

descriptive statistics of $t = \dfrac{\hat{\bar{y}}_R - \bar{y}}{\sqrt{v}}$ based on 1000 simple random samples of $n = 32$

Population	variance estimator	nominal coverage probability $(1-\alpha) \times 100\%$					bias	s.d. (of t)	coefficient of skewness
		99	95	90	80	70			
1	v_0	97.1	91.5	84.8	74.5	64.6	-0.011	1.23	0.066
	v_1	97.7	92.1	86.3	75.2	64.6	-0.012	1.18	0.035
	v_2	97.8	93.0	87.1	76.0	64.7	-0.012	1.16	0.030
	$v_{g_{opt}}$	97.3	92.6	86.5	75.0	64.7	-0.009	1.18	0.051
	$v_{\tilde{g}}$	97.7	92.9	86.7	75.0	64.6	-0.006	1.19	0.049
	v_{reg}	97.7	92.5	86.7	75.1	64.6	-0.009	1.18	0.040
	v_H	97.8	93.2	87.8	76.7	65.5	-0.012	1.14	0.030
	v_D	97.8	93.4	88.0	77.2	66.1	-0.012	1.13	0.034
	v_J	97.9	94.1	88.6	78.6	67.7	-0.013	1.10	0.033

Population	variance estimator	nominal coverage probability $(1-\alpha) \times 100\%$					bias	s.d. (of t)	coefficient of skewness
		99	95	90	80	70			
2	v_0	96.3	90.4	85.7	75.9	67.5	-0.17	1.22	-0.78
	v_1	96.1	91.0	86.0	76.8	68.1	-0.18	1.21	-0.83
	v_2	95.7	91.4	86.4	76.4	68.2	-0.19	1.20	-0.87
	$v_{\hat{g}_{opt}}$	96.0	90.7	85.7	76.1	67.6	-0.18	1.21	-0.81
	$v_{\tilde{g}}$	95.9	91.1	85.8	76.2	67.4	-0.18	1.21	-0.81
	v_{reg}	95.9	91.0	85.9	76.6	67.8	-0.18	1.21	-0.82
	v_H	95.7	91.8	86.6	77.2	68.7	-0.19	1.19	-0.88
	v_D	95.7	91.6	86.5	77.1	68.6	-0.19	1.19	-0.86
	v_J	96.0	92.3	87.0	77.9	68.8	-0.18	1.17	-0.88
3	v_0	93.1	86.7	80.7	71.2	63.4	0.23	1.42	1.02
	v_1	94.6	88.8	80.9	71.6	62.9	0.13	1.31	0.57
	v_2	96.9	88.2	82.5	71.1	62.3	0.04	1.25	0.13
	$v_{\hat{g}_{opt}}$	96.2	90.0	83.5	73.1	64.6	0.12	1.24	0.55
	$v_{\tilde{g}}$	96.5	89.1	82.4	72.2	63.1	0.08	1.25	0.41
	v_{reg}	95.2	88.7	81.1	71.1	62.6	0.11	1.30	0.51
	v_H	97.4	88.8	83.1	72.0	62.7	0.03	1.23	0.08
	v_D	97.5	89.9	83.7	73.3	64.2	0.04	1.21	0.11
	v_J	97.8	91.2	85.6	75.2	66.3	0.05	1.16	0.21

4								
v_0	85.8	77.1	72.1	63.7	55.9	-0.79	1.69	-0.95
v_1	87.3	77.7	72.6	63.7	55.0	-0.74	1.59	-0.79
v_2	88.6	78.9	72.3	63.9	54.6	-0.69	1.53	-0.71
$v_{\hat{g}_{opt}}$	87.5	77.8	72.4	63.2	55.2	-0.74	1.60	-0.82
$v_{\tilde{g}}$	87.3	77.9	72.1	63.3	55.0	-0.73	1.59	-0.78
v_{reg}	87.0	77.9	72.2	63.3	54.7	-0.73	1.59	-0.77
v_H	89.0	79.9	73.1	64.5	55.2	-0.68	1.51	-0.71
v_D	89.7	79.9	73.9	65.4	55.7	-0.68	1.48	-0.74
v_J	90.6	81.5	75.4	66.5	57.9	-0.66	1.43	-0.77
5								
v_0	97.3	92.9	88.0	77.5	67.3	0.12	1.14	0.40
v_1	97.8	93.4	89.2	77.9	67.5	0.09	1.11	0.27
v_2	97.9	94.0	90.1	77.5	67.7	0.06	1.10	0.15
$v_{\hat{g}_{opt}}$	97.9	93.7	89.4	77.3	67.5	0.08	1.11	0.23
$v_{\tilde{g}}$	97.8	93.5	89.2	77.7	67.6	0.07	1.11	0.21
v_{reg}	97.8	93.4	89.4	77.9	67.6	0.08	1.11	0.25
v_H	98.2	94.3	90.3	78.6	68.3	0.06	1.08	0.14
v_D	98.1	94.3	90.1	79.1	68.3	0.06	1.08	0.14
v_J	98.2	94.6	90.9	78.4	68.9	0.06	1.07	0.16

Population	variance estimator	nominal coverage probability $(1-\alpha)\times 100\%$					bias	s.d. (of t)	coefficient of skewness
		99	95	90	80	70			
6	v_0	97.3	92.2	84.3	73.4	65.9	-0.001	1.19	0.055
	v_1	98.6	93.3	86.4	75.5	67.1	0.001	1.13	0.023
	v_2	98.4	93.4	87.6	78.0	67.8	0.021	1.10	0.066
	$v_{\hat{g}opt}$	97.8	93.3	86.2	75.6	68.3	0.011	1.14	0.062
	$v_{\tilde{g}}$	97.9	93.8	86.1	76.1	67.7	0.021	1.13	0.075
	v_{reg}	98.3	93.8	86.2	75.4	67.5	0.013	1.13	0.035
	v_H	98.5	93.8	87.9	79.1	68.9	0.023	1.08	0.078
	v_D	98.8	94.6	88.4	80.0	70.6	0.020	1.06	0.080
	v_J	99.1	95.8	89.8	81.7	72.0	0.015	1.01	0.072
7	v_0	95.7	91.8	87.8	79.6	70.5	0.24	1.23	1.46
	v_1	97.5	92.7	88.2	80.0	70.5	0.13	1.12	0.63
	v_2	98.4	93.7	88.4	80.1	70.1	0.03	1.08	-0.11
	$v_{\hat{g}opt}$	95.7	92.1	88.7	81.5	71.6	0.25	1.21	1.60
	$v_{\tilde{g}}$	99.0	94.5	89.7	81.2	71.7	0.06	1.04	0.16
	v_{reg}	98.1	93.2	88.4	80.4	70.6	0.10	1.09	0.47
	v_H	98.3	94.0	89.2	80.5	71.3	0.01	1.06	-0.19
	v_D	98.6	94.2	89.5	81.0	72.3	0.01	1.04	-0.20
	v_J	98.7	94.7	90.7	82.8	73.6	0.02	1.01	-0.11

8

v_0	95.7	87.1	78.1	64.5	53.3	-0.25	1.36	-0.01
v_1	97.2	88.9	79.7	65.5	53.2	-0.23	1.31	0.02
v_2	97.5	88.6	80.8	66.8	52.8	-0.21	1.28	0.06
$v_{\hat{g}_{opt}}$	97.4	89.3	80.7	67.1	53.8	-0.21	1.28	0.01
$v_{\tilde{g}}$	97.6	89.1	80.5	67.9	54.5	-0.20	1.27	0.06
v_{reg}	96.8	87.8	79.4	66.5	52.5	-0.22	1.31	-0.05
v_H	97.9	88.7	81.6	67.8	53.9	-0.20	1.26	0.06
v_D	98.3	90.4	84.0	69.8	55.6	-0.20	1.21	0.05
v_J	98.7	93.6	81.6	67.8	53.9	-0.20	1.15	0.03

9

v_0	91.2	86.3	82.5	75.3	67.3	0.50	1.67	2.40
v_1	93.1	86.9	82.6	74.9	66.5	0.36	1.46	1.57
v_2	94.6	88.1	82.2	74.2	66.6	0.25	1.32	0.88
$v_{\hat{g}_{opt}}$	93.1	87.8	83.7	76.5	68.4	0.38	1.40	1.64
$v_{\tilde{g}}$	96.2	89.5	83.7	74.6	67.0	0.22	1.26	0.87
v_{reg}	93.8	88.0	82.5	74.4	66.7	0.30	1.39	1.34
v_H	94.9	88.7	82.6	74.6	67.6	0.23	1.30	0.81
v_D	95.4	89.2	83.7	75.4	68.9	0.23	1.27	0.85
v_J	95.7	90.5	85.5	76.8	70.3	0.23	1.23	0.96

discrepancy is most serious in population 4 where the
t-statistics are "abnormal", and is serious in population 8
for $1 - \alpha = 0.7, 0.8, 0.9$.

(ii) The estimator v_0 is least reliable in that the
actual coverage probability is far. off the nominal level.
The most reliable one among the nine estimators is the
jackknife v_J with v_D as the second best. On populations
1 and 5-9 v_D performs as well or nearly as well as v_J.
The performance of v_H is comparable to v_D (and
sometimes v_J) on populations 1, 2, 4, 5 and 7. Other
estimators v_1, v_2, $v_{\hat{g}}$, $v_{\tilde{g}}$, v_{reg} are comparable among each
other but trail slightly behind v_H.

(iii) Among v_0, v_1, v_2, v_2 is the best, v_1 the
middle and v_0 the worst for most cases. A partial
explanation is that v_2 is the only one among the three
that is asymptotically equivalent to v_J, the best
performer.

The excellent performance of v_J might be explained by
the large expected value Ev_J, or equivalently the width of
the associated t-interval. In the same spirit, could the
better performance of v_2 relative to v_1 and v_1 to
v_0 be attributed to any similar behavior in their
t-intervals? As remarked in II they are all short
intervals, but from Table 3 the biases of v_0, v_1, v_2 do
not exhibit any clear-cut pattern to support such claim. We
do not believe that the length of the interval alone
explains the difference.

One obvious thing to observe from comparing Tables 3
and 4 is that, estimators like $v_{\hat{g}}$, $v_{\tilde{g}}$, v_{reg} that perform
well for estimating $MSE(\hat{\bar{y}}_R)$ do not fare very well for
giving reliable interval estimate. On the other hand,
v_J, though having very large mean square error for
estimating $MSE(\hat{\bar{y}}_R)$, is extremely good in giving reliable
interval estimate. Perhaps the only consistent conclusion
from Table 3 and 4 is that v_0 fares poorly in both
criteria. Since a variance estimator is primarily judged by
the quality of the associated interval estimator, an
important question thus arises: What properties of v as a

point estimator will provide a good guide in judging its
performance as an interval estimator?

We propose to take $\bar{x}$ as an <u>ancillary</u> statistic and
draw inference by conditioning on $\bar{x}$. More precisely an
estimate of the conditional mean square error $MSE(\hat{\bar{y}}_R | \bar{x})$
should be used in the interval estimate of $\bar{Y}$. In the
context of maximum likelihood estimation Efron and Hinkley
(1978) proposed a version of conditional variance (given an
appropriate ancillary statistic) for constructing reliable
interval estimate.

To see how different variance estimators perform in
tracking $MSE(\hat{\bar{y}}_R | \bar{x})$, we divide the 1000 samples into 20
groups of 50 samples according to the order of the $\bar{x}$
values. For each group we calculate the average of $\bar{x}$,
$\sum_1^{50} \bar{x}/50$, the conditional MSE of $\hat{\bar{y}}_R$ within the
group, $\sum_1^{50} (\hat{\bar{y}}_R - \bar{Y})^2/50$ and the averages of each of the
nine estimates, $\bar{v}_0 = \sum_1^{50} v_0/50$, etc. We then plot the
values of $N\sqrt{MSE}$, $N\sqrt{\bar{v}_0}$, $N\sqrt{\bar{v}_1}$, and so on, against the
average values of $\bar{x}$ (the factor N is to make our plots
comparable to those of Royall and Cumberland (1981).) To
save space we only show the plots for populations 2 and 3 in
Figures 5 and 6. Plots for other populations exhibit
similar patterns. In Figures 5 and 6, the trajectories of
$\sqrt{\bar{v}_{\tilde{g}}}$, $\sqrt{\bar{v}_{reg}}$, $\sqrt{\bar{v}_D}$ are omitted because they are too close to
the trajectories of $\sqrt{\bar{v}_{\hat{g}}}$ and $\sqrt{\bar{v}_H}$ respectively. One can
see that $\sqrt{\bar{v}_H}$, $\sqrt{\bar{v}_J}$ and $\sqrt{\bar{v}_2}$ seem to track the conditional
$\sqrt{MSE}$ (trajectory) better than $\sqrt{\bar{v}_0}$, $\sqrt{\bar{v}_1}$ and $\sqrt{\bar{v}_{\hat{g}}}$. Such a
visual comparison of trajectories is somewhat arbitrary and
imprecise. Instead we consider a measure of distance (22)
between the $\sqrt{MSE}$-trajectory and any $\sqrt{\bar{v}}$-trajectory given by

$$\{\frac{1}{20} \sum_1^{20} (\sqrt{\bar{v}} - \sqrt{MSE})^2\}^{1/2} , \tag{22}$$

where the summation is over the twenty groups of values. If
the distance measure for a variance estimator v is
smaller, we say that v is closer to the conditional MSE of

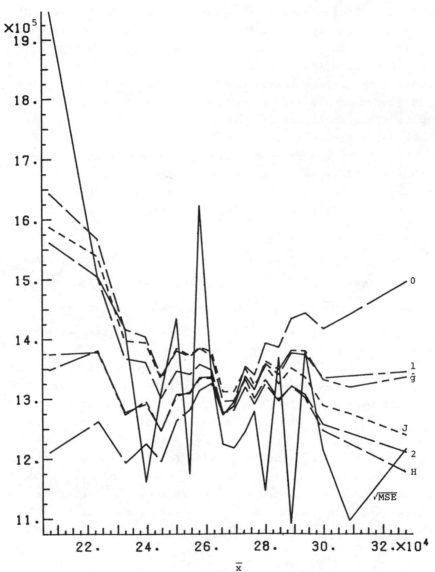

Figure 5. Curves $\sqrt{MSE}$, $\sqrt{v_0}$, $\sqrt{v_1}$, $\sqrt{v_2}$, $\sqrt{v_{\hat{g}}}$, $\sqrt{v_J}$, $\sqrt{v_H}$ for population (2).

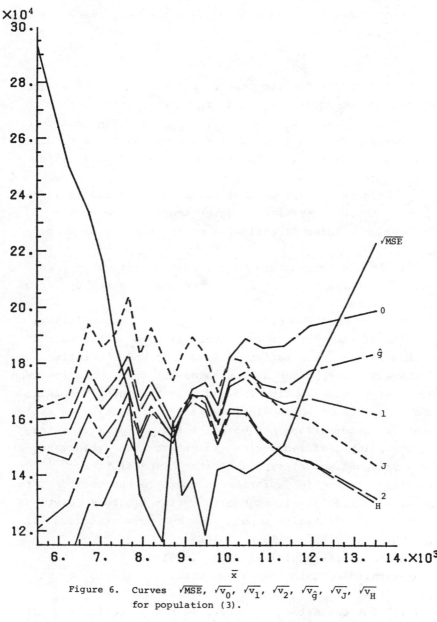

Figure 6.   Curves $\sqrt{\text{MSE}}$, $\sqrt{v_0}$, $\sqrt{v_1}$, $\sqrt{v_2}$, $\sqrt{v_{\hat{g}}}$, $\sqrt{v_J}$, $\sqrt{v_H}$
            for population (3).

the ratio estimator. In Table 5 we list such values for
nine populations and nine estimators.

From Table 5 we can roughly rank the performance of the
nine estimators as

$$v_H, \ v_D > v_J, \ v_2, \ v_{\tilde{g}} > v_{\hat{g}}, \ v_{reg} > v_1 > v_0 \ ,$$

where ">" means "better than". Again v_0 is the worst,
v_1 the second worst. v_H and v_D are slightly better
than v_J, v_2 and $v_{\tilde{g}}$. And $v_{\hat{g}}$, v_{reg} are the mediocre
performers. This is in general agreement with the results
of Table 4. We are thus led to the tentative conclusion
that

> "Variance estimators that estimate the conditional MSE
> of the ratio estimator better tend to give more
> reliable interval estimates of the population mean."

It should be possible to justify theoretically this
statement at least by assuming a reasonable superpopulation
model between y and x.

The apparent contradiction between the unconditional
behavior of the variance estimators as <u>point</u> estimators of
the MSE of the ratio estimator (Table 3) and the reliability
of the associated <u>interval</u> estimators of the population mean
(Table 4) is now happily resolved. If the inference is made
conditional on an appropriate ancillary statistic, in this
case $\bar{x}$, good performance of an estimator for estimating
the conditional variance often points to good performance of
the corresponding interval estimator. Therefore our
empirical finding lends further support to the work of Efron
and Hinkley (1978), although ours is for finite populations
and theirs for infinite populations and parametric models.

6. <u>CONCLUSIONS AND FURTHER REMARKS</u>

Based on the empirical study in §4 and §5 and the
theoretical discussion in §3, we arrive at the following
conclusions.

1) The estimator v_0, (4), is the poorest among the
nine estimators considered in the paper. Its t-intervals
are not reliable and it does not estimate either the MSE or
the conditional MSE of $\hat{\bar{y}}_R$ well. However it is the most

Table 5. Distance, formula (22), between $\sqrt{MSE}$-trajectory and $\sqrt{v}$-trajectory

variance estimator	Population								
	1	2	3	4	5	6	7	8	9
v_0	0.528	1998	227	339	8.90	15888	5.62	1.60	3.79
v_1	0.354	1535	197	225	6.85	9916	5.31	1.31	3.55
v_2	0.254	1170	171	184	5.30	7477	5.02	1.21	3.32
v_{reg}	0.325	1532	186	200	6.45	8933	5.19	1.21	3.40
$v_{\hat{g}_{opt}}$	0.320	1559	172	181	6.05	9837	5.54	1.16	3.42
$v_{\tilde{g}}$	0.297	1495	160	176	6.01	9820	4.91	1.13	3.05
v_H	0.241	1118	168	176	5.20	7579	5.04	1.14	3.32
v_D	0.232	1118	170	155	5.38	7902	5.36	0.96	3.39
v_J	0.240	1221	179	166	5.20	10181	5.15	0.81	3.55
unit	10^0	10^6	10^4	10^4	10^0	10^{14}	10^0	10^0	10^0

commonly recommended estimator in virtually every textbook on sampling. (In fact some well-known textbooks do not even mention the better estimator v_2.) The "de-mystification" of v_0 is probably the most useful of all the recommendations made in our paper.

2) Among v_0, v_1, v_2, v_2 is better than v_1 and v_1 better than v_0 for giving reliable t-intervals. The performance of v_0, v_1 and v_2 for estimating MSE depends on the underlying populations and has no direct bearing on the performance of interval estimates.

3) If more complicated computations are allowed (such may be an issue for large scale surveys), we have more choices. The jackknife v_J gives very reliable t-intervals and v_H, v_D are almost as good. Note that for large samples, all three estimators are close to v_2, but not to any other v_g, $g \neq 2$. The reason that v_J does so poorly for estimating MSE is because it estimates the conditional MSE well, and typically the conditional MSE varies greatly with $\bar{x}$. This instability of v_J for estimating the unconditional MSE has also been reported in previous papers but should not concern us any more.

4) The estimators $v_{\hat{g}}$, $v_{\tilde{g}}$ and v_{reg} are asymptotically equivalent. They are good for estimating the unconditional MSE but are mediocre for giving reliable t-intervals.

We emphasize that reliable t-intervals seems to be related to the good performance of v for estimating the conditional MSE. The problem of choosing a proper ancillary statistic and making inference conditional on it is an important one in the theory of survey sampling.

Encouraged by the relatively good performance of v_2 over v_0, we have considered the variance estimation problem in other settings. For the regression estimator under simple random sampling,

$$\bar{y}_{\ell r} = \bar{y} + b(\bar{X} - \bar{x}), \quad b = \frac{s_{xy}}{s_x^2}$$

the typical estimator of $\text{Var}(\bar{y}_{\ell r})$ is (Cochran, 1977, p. 195)

$$\widehat{var} = \frac{1 - f}{n} \frac{1}{n - 2} \sum_{1}^{n} (y_i - \bar{y} - b(x_i - \bar{x}))^2 ,$$

which is the sample analogue of the approximate variance of $\bar{y}_{\ell r}$. It is natural to consider the following class of estimators

$$(\frac{\bar{X}}{\bar{x}})^g \widehat{var}, \quad \text{especially} \quad g = 2 .$$

A detailed report will be available later.

In stratified random sampling with small sample size per stratum, the combined ratio estimator is often used. Let $W_h = N_h/N$ be the hth stratum weight, $\bar{y}_h, \bar{x}_h, \bar{Y}_h, \bar{X}_h$ be the y- and x-sample and population means of the hth stratum. The combined ratio estimator is

$$\hat{\bar{y}}_{RC} = \frac{\bar{y}_{st}}{\bar{x}_{st}} \bar{X}, \quad \bar{y}_{st} = \sum_{1}^{L} W_h \bar{y}_h, \quad \bar{x}_{st} = \sum_{1}^{L} W_h \bar{x}_h .$$

Its approximate variance is

$$\sum_{1}^{L} \frac{1 - f_h}{n_h} \frac{1}{N_h - 1} \sum_{1}^{N_h} (y_{hi} - \bar{Y}_h - \frac{\bar{Y}}{\bar{X}} (x_{hi} - \bar{X}_h))^2 .$$

The following class of estimators

$$(\frac{\bar{X}}{\bar{x}_{st}})^g \sum_{1}^{L} \frac{1 - f_h}{n_h} \frac{1}{n_h - 1} \sum_{1}^{n_h} (y_{hi} - \bar{y}_h - \frac{\bar{y}_{st}}{\bar{x}_{st}} (x_{hi} - \bar{x}_h))^2$$

has been considered by the first author (CFW). The case $g = 2$ is of special interest. The detailed results will be reported elsewhere.

Extensions in other situations are obvious.

REFERENCES

1. Brewer, K. R. W., Ratio estimation and finite populations: Some results deducible from the assumption of an underlying stochastic process, Australian Journal of Statistics 5 (1963), 93-105.

2. Cochran, W. G., Sampling Techniques, 3rd Edition, Wiley, New York, 1977.

3. Efron, B. and D. V. Hinkley, Assessing the accuracy of the maximum likelihood estimator: observed versus expected Fisher information, Biometrika 65 (1978), 457-487.

4. Fuller, W. A., A note on regression estimation for sample surveys, unpublished manuscript, 1977.

5. _____, Comment on a paper by Royall and Cumberland, Journal of the American Statistical Association 76 (1981), 78-80.

6. Krewski, D. and R. P. Chakrabarty, On the stability of the jackknife variance estimator in ratio estimation, Journal of Statistical Planning and Inference 5 (1981), 71-78.

7. Rao, J. N. K., Some small sample results in ratio and regression estimation, Journal of the Indian Statistical Association 6 (1968), 160-168.

8. _____, Ratio and regression estimators, in New Developments in Survey Sampling (N. L. John and H. Smith, eds.), Wley, New York, 1969, 213-234.

9. _____, Sampling designs involving unequal probabilities of selection and robust estimation of a finite population total, in Contributions to Survey Sampling and Applied Statistics (H. A. David, ed.), Academic Press, New York, 1978, 69-87.

10. _____ and L. D. Beegle, A Monte Carlo study of some ratio estimators, Sankhya B 29 (1967), 47-56.

11. _____ and R. A. Kuzik, Sampling errors in ratio estimation, Sankhya C 36 (1974), 43-58.

12. Rao, P. S. R. S. and J.N. K. Rao, Small sample results for ratio estimators, Biometrika 58 (1971), 625-630.

13. Royall, R. M., On finite population sampling theory
     under certain linear regression models, Biometrika 57
     (1970), 377-387.

14. _____ and W. G. Cumberland, Variance
     estimation in finite population sampling, Journal of
     the American Statistical Association 73 (1978),
     351-358.

15. _____ and _____, An empirical study
     of prediction theory in finite population sampling:
     simple random samping and the ratio estimator, in
     Survey Sampling and Measurements (N.K. Namboodiri,
     ed.), Academic Press, New York, 1978, 293-309.

16. _____ and _____, An empirical study
     of the ratio estimator and estimators of its variance,
     Journal of the American Statistical Association 76
     (1981), 66-77.

17. _____ and K. R. Eberhardt, Variance estimates
     for the ratio estimator, Sankhya C 37 (1975), 43-52.

18. Wu, C. F., Estimation of variance of the ratio
     estimator, Biometrika 69 (1982), 183-189.

The research was supported by National Science Foundation
Grant No. MCS-7901846 and sponsored by the U. S. Army under
Contract No. DAAG29-80-C-0041.

                         Mathematics Research Center
                         and Department of Statistics
                         University of Wisconsin-Madison
                         Madison, WI 53706

                         Department of Statistics
                         University of Wisconsin-Madison
                         Madison, WI 53706

Autocorrelation-robust Design
of Experiments

J. Kiefer and H. P. Wynn

1. BACKGROUND

There has been considerable recent research interest in designing experiments under error processes which have spatial autocorrelation. The classical paper on the subject is Williams (1952) which was extended by Kiefer (1960).

The basic model is the linear model

$$E(Y) = X\alpha, \qquad Cov(Y) = V \tag{1}$$

where the parameter vector α or some special set of contrasts θ, such as all treatment contrasts, is to be estimated. The matrix V derives from some autocorrelated process whose geometry is related to the design region and hence to the design matrix X. The published work in the field divides roughly into the following approaches.

(i)   V is known and we look for optimum designs for the ordinary least squares estimate $\hat{\theta}_0$ of θ but under model (1), above.

(ii)   V is known and we look for optimum designs but use the best linear unbiased estimate $\hat{\theta}_V$ under (1).

(iii)   V is known and investigate designs for which $\hat{\theta}_0 \equiv \hat{\theta}_V$.

(iv)   V is unknown and consider design and analysis.

(v)   V is unknown but look for designs robust against a range of possible V matrices, e.g. minimax.

Thus (i) above was treated by Kiefer and Wynn (1980) under a simple neighbor model for V. That paper also contains an extensive list of references. Approaches (ii) and (v) were taken by Kiefer and Wynn (1981). Some study of (iii) is carried out by Martin (1981).

2. ONE DIMENSION: ANALYSIS

The paper by Kiefer and Wynn (1981) gives a thorough foundation for optimum treatment design in one dimension in the presence of autocorrelated errors. A brief resumé is necessary to provide a statistical background for the new combinatorial constructions of this paper.

Let $\{Y_t\}$ be a p-th order stationary autoregressive process on the line in discrete time. Thus

$$Y_t + \rho_1 Y_{r-1} + \cdots + \rho_p Y_{r-p} = \varepsilon_t$$

($t = 0,1,2,\ldots$.), where $\{\varepsilon_t\}$ is an uncorrelated zero mean innovation process with variance σ^2. The actual <u>observation</u> at time t is

$$W_t = Y_t + \alpha_{[t]}$$

where $\alpha_{[t]} \in \{\alpha_1,\ldots,\alpha_k\}$ is the unknown parameter for one of k treatments allocated at time t. We are interested in estimating the symmetric contrast vector

$$\theta = (I - \frac{1}{k} J)\alpha \quad ,$$

where I is the k×k identity and J the k×k matrix of ones. We assume that $\rho = (\rho_1,\ldots,\rho_p)$ is known and work with the best linear unbiased estimate $\hat{\theta}$ of θ (under finite sample size N). The design optimality criterion is to minimize, through the choice of treatment allocation to time points the quantity

$$\Phi \{cov(\hat{\theta})^-\} \tag{2}$$

where A^- is the Moore-Penrose g-inverse of a matrix A, $cov(\hat{\theta})$ is the covariance of $\hat{\theta}$ under the model and Φ is a member of a special class of functionals on k×k non-negative definite matrices. This class comprises all Φ which are (i) convex, (ii) invariant under permutations of coordinates and (iii) non-increasing: G-H non-negative definite => $\Phi(G) \le \Phi(H)$. If a design minimizes (2) for all such Φ (fixed ρ) then it is said to be universally optimum.

Let $C = \sigma^2(cov(\hat{\theta}))^-$. Then sufficient conditions for universal optimality can be given in terms of the form of C. These derive from Proposition 1 of Kiefer (1975). They are

(a) C is completely symmetric: i.e. C is a multiply of $I - \frac{1}{k} J$.

(b) trace (C) is a maximum.

(Note, this is a simpler version than in Kiefer and Wynn, 1981, which is allowed if we assume equal treatment replication.)

To realize (a) and (b) we need to convert them into conditions on the design.

It turns out that all that is needed is the adjacencies counts (or proportions), namely the number of times treatments are next to themselves up to lag p (where p, we recall, is the order of the process).

To make the description easier to understand label the treatments A,B,C... Define

$$\pi_{r,N} = \frac{1}{N} \{\text{number of times A is r steps from A, or B is r-steps from B,....}\} \qquad (r = 1,...,p)$$

For example consider the sequence

ABAABACAABBCABCA   ,

with N = 16. If p = 3 we calculate

$$\pi_{1,N} = \frac{3}{16} , \quad \pi_{2,N} = \frac{3}{16} , \quad \pi_{3,N} = \frac{3}{8} .$$

It is very convenient to take an asymptotic approach. Thus we call a sequence allowable if

$$\pi_{r,N} \rightarrow \text{a limit,} \quad \pi_r$$

as $n \rightarrow \infty$. With this notation we can evaluate an asymptotic version of trace (C) which is

$$\frac{1}{N} \text{trace} (C) = \sum_{r=0}^{p} \rho_r^2 - \frac{1}{k} (\sum_{r=0}^{p} \rho_r)^2$$
$$+ 2 \sum_{r=1}^{p} \pi_r \sum_{s=0}^{p-r} \rho_s \rho_{s+r} .$$

Condition (b) says maximize

$$\sum_{r=1}^{p} \pi_r \sum_{s=0}^{p-r} \rho_s \rho_{s+r} \qquad (3).$$

Condition (a) is achieved by "complete symmetrization"
namely taking sequences in which the adjacency counts
between different treatments (up to lag p) are unaltered by a
permutation of letters (treatment labels).

Define for k treatments

$$\Pi_p(k) = \{(\pi_1,\ldots,\pi_p)\}$$

that is, the set of $(\pi_1,\ldots,\pi_p)$ vectors attainable in the
limit. It can be shown, and is fully discussed in the next
section, that $\Pi_p(k)$ is a polyhedron whose vertices are
obtained from purely periodic sequences with finite period.
Since (3) is linear in the π_r it is maximised at a vertex and
thus if the extreme periodic sequences can be found one of
them will be optimum for a given ρ. In principle, then, the
optimisation problem is solved. For example, with k = 2 and
p = 3 the extreme π-vectors with their sequences are:

Sequence	π-vector
A . . .	(1, 1, 1)
AB . . .	(0, 1, 0)
AAB . . .	$(\frac{1}{3}, \frac{1}{3}, 1)$
AABB . . .	$(\frac{1}{2}, 0, \frac{1}{2})$
AAABBB . . .	$(\frac{2}{3}, \frac{1}{3}, 0)$

Just a full cycle is given above, it being assumed that the
cycle is repeated indefinitely.

The extreme sequences for k = 2, p = 2 are A (1,1),
AB (0,1) and AABB $(\frac{1}{2}, 0)$. For p = 3 and k = 3 and ≥ 4 are
given in Kiefer and Wynn (1981). A theorem, there, says that
all $\Pi_p(k)$ are identical for k $\geq$ p+1.

The next two sections develop the theory of Π-regions for
one and more dimensions.

3.   STATIONARY BINARY PROCESSES IN ONE DIMENSION

We shall restrict ourselves to designs with 2 treatments
k = 2. For the moment label A $\equiv$ 1 and B $\equiv$ -1. Then the
design can be written as a real sequence $\ldots X_{-1}, X_0, X_1, \ldots$
with $X_t = \pm 1$. Then

$$2\pi_r - 1 = \frac{1}{N} \sum_{r=p}^{N} X_r X_{r-p} \tag{4}$$

If we subtract $(\frac{1}{N} \overset{.}{\underset{r=p}{\overset{N}{\Sigma}}} X_r)^2$ we have the sample autocovariance
function $c(r)$ of the sequence $\{X_t\}$. Assume that $\lim \frac{1}{N} \Sigma X_r = 0$.
Then the condition that $\pi_{r,N} \to$ a limit π_r $(r = 1, \ldots, p)$ is
then a statement that the sample autocovariance function $c(r)$
converges $(r = 1, \ldots, p)$. Now suppose that the binary sequence
$\{X_t\}$ is a single realization of a stationary ergodic zero mean
binary process then (with probability one) the sample autocovariance function will converge to the actual autocovariance
function, by the Ergodic Theorem. The key fact is that the
converse holds: for any $c(r) = 2\pi_r - 1$ with $\pi_r \in \Pi_p(2)$ there is
a stationary ergodic binary process with $c(r)$ as its autocovariance function. Thus there is an exact one-one
correspondence between the $c(r)$ and π_r.

There is a limited literature in this area. Hobby and
Ylvisaker (1964) look at stationary discrete time finite state
processes. There are some unpublished notes of the late
Walter Weissblum (Bell Telephone Laboratories). The early
study of covariances for binary processes is due to Shepp
(1967), see also Masry (1972). We have also used the work of
Martins de Carvalho and Clark (1977, 1981).

It is essentially easier to find the extreme distributions
among stationary binary distributions on the line rather than
the extreme covariances. It is not true that every extreme
distribution gives an extreme covariance. The problem of
finding a characterisation of extreme distributions was essentially solved by Hobby and Ylvisaker (1964) and given a
precise formulation in the unpublished notes of Weissblum
(Zaman, 1981, has also exploited this work independently of
the authors in the study of Markov exchangeable sequences).
We now explain this theory.

Let $X_0, X_1, X_2, \ldots$ be a stationary binary process. Consider
a window of length p+1 covering p+1 adjacent time points, say
$t = 0, 1, \ldots, p+1$. The process induces distributions at any
finite set of points. So define

$$P_{p+1}(x_0, \ldots, x_p) = \text{Prob}(X_0 = x_1, \ldots, X_p = x_p)$$

Then

$$P_p(x_1, \ldots, x_p) = \text{Prob}(X_1 = x_1, \ldots, X_p = x_p)$$

can be expressed in two ways

$$P_p(x_1,\ldots,x_p) = \sum_{x=0,1} P(x,x_1,\ldots,x_p) = \sum_{x=0,1} P(x_1,\ldots,x_p,x)$$

(5)

Thus we get a bank of 2^p equations (5) one for each p-tple $(x_1,\ldots,x_p)$. It is notationally convenient in writing out examples to drop the P notation and just write $P(x_1,\ldots,x_p) \equiv x_0 \ldots\ldots x_p$ identifying the probability with the binary string itself. Then, for example, for p = 2 we obtain the 4 equations

```
0 0 0  +  1 0 0  =  0 0 0  +  0 0 1
0 0 1  +  1 0 1  =  0 1 0  +  0 1 1
0 1 0  +  1 1 0  =  1 0 0  +  1 0 1
0 1 1  +  1 1 1  =  1 1 0  +  1 1 1
```
(6)

By a "solution" to (5) we mean any collection of $P(x_0,\ldots,x_p) \geq 0$ with $\Sigma P(x_0,\ldots,x_p) = 1$ satisfying the equations. Any such solution is a distribution P on the vertices of the 2^{p+1} unit cube $\bigotimes_{p+1} [0,1]$. The Hobby-Ylvisaker-Weissblum theorem (H-Y-W) then characterises the extreme distributions among all those satisfying (5).

Theorem 1 (H-Y-W). The extreme distributions in the (p+1)-tples amongst all those obtained from stationary binary sequences are precisely all those for periodic sequences whose cycles do not repeat a p-tple (together with a uniform random phase shift).

Proof. We sketch the main ideas of the proof using the p = 2 example. Let $\underline{P}$ be a solution to (5). Select one equation for which on the left hand side $P(x_0,\ldots,x_p) > 0$. Label $x^{(1)} = (x_0,\ldots,x_p)$ and $P(x_0,\ldots,x_p) = \alpha_1 > 0$. Then on the right hand side of this equation there must be an $x^{(2)} = (x_1,\ldots,x_p,x')$ with $P(x_1,\ldots,x_p,x') = \alpha_2 > 0$. Then find $x^{(2)}$ again in one (possibly new) equation on the LHS. This means in turn that there is an $x^{(3)} = (x_2,\ldots,x_p,x',x'')$ on the RHS of this (new) equation with $P(x_2,\ldots,x_p,x',x'') = \alpha_3 > 0$. Continue in this way until we have returned to the same equation for the first time. Then we have a cycle

$$x^{(r)}, x^{(r+1)}, \ldots, x^{(r+\ell_1)}$$

with $x^{(r)}$ and $x^{(r+\ell_1)}$ in the same equation (on LHS). Let

$$\alpha^{(1)} = \min_{1 \leq s < \ell_1} \alpha_{r+s} \quad .$$

Then "extract" this cycle. That is, express the solution $\underset{\sim}{P}$ as

$$\underset{\sim}{P} = \alpha^{(1)}\ell_1 \ \underset{\sim}{P}_1 + (1 - \alpha^{(1)}\ell_1)\underset{\sim}{P}_1'$$

where $\underset{\sim}{P}_1$ is the distribution which attaches mass $\frac{1}{\ell_1}$ to each of the (p+1)-tples $x^{r+1}, \ldots, x^{r+\ell_1}$. Now clearly the "residual" $\underset{\sim}{P}_1'$ also satisfies the equation (5). Thus we may extract another cycle $\underset{\sim}{P}_2$ and write

$$\underset{\sim}{P}_1' = \alpha^{(2)}\ell_2 \ \underset{\sim}{P}_2 + (1 - \alpha^{(2)}\ell_2)\underset{\sim}{P}_2'$$

and so on. Thus P is decomposed into cycles

$$\underset{\sim}{P} = \sum_{i=1}^{M} \beta_i \ \underset{\sim}{P}_i$$

where $\beta_i \geq 0 \ \Sigma\beta_i = 1$. It is straightforward to show that every such cycle $\underset{\sim}{P}_i$ is extreme. Then, finally, we see that every $\underset{\sim}{P}_i$ is realizable as the (p+1)-window distribution from a stationary ergodic binary sequence. Stringing together the (p+1)-tples we get a cycle of x_i (= 0 or 1) of length ℓ. Take the process which is obtained by taking with probability $\frac{1}{\ell}$ each of the ℓ possible phase positions of the strictly period process formed by repeating the x_i cycle indefinitely. This process has the required $\underset{\sim}{P}_i$. The defining property of the $\underset{\sim}{P}_i$ cycles is that they do not repeat p-tples. Thus we have Theorem 1.

The theorem is close to theorems in transport theory where the (p+1)-tples are nodes and the $P(x_0, \ldots, x_p)$ are flows.

To make Theorem 1 clear consider the p = 2 case. The possible cycles in Theorem 1 are, referring to (6):

$$
\begin{array}{rl}
\text{(i)} & 0\ 0\ 0 \rightarrow (0\ 0\ 0) \\
\text{(ii)} & 1\ 1\ 1 \rightarrow (1\ 1\ 1) \\
\text{(iii)} & 0\ 1\ 0 \rightarrow 1\ 0\ 1 \rightarrow (0\ 1\ 0) \\
\text{(iv)} & 0\ 0\ 1 \rightarrow 0\ 1\ 0 \rightarrow 1\ 0\ 0 \rightarrow (0\ 0\ 1) \\
\text{(v)} & 1\ 1\ 0 \rightarrow 1\ 0\ 1 \rightarrow 0\ 1\ 1 \rightarrow (1\ 1\ 0) \\
\text{(vi)} & 0\ 0\ 1 \rightarrow 0\ 1\ 1 \rightarrow 1\ 1\ 0 \rightarrow 1\ 0\ 0 \rightarrow (0\ 0\ 1) \qquad (7)
\end{array}
$$

The extreme periodic sequences are (i) 0... (ii) 1...
(iii) 01... (iv) 001... (v) 110... (vi) 0011... . These have
π-vectors (p = 2) (i) and (ii) (1,1),... (iii) (0,1), (iv) and
(v) $(\frac{1}{3}, \frac{1}{3})$ and (vi) $(\frac{1}{2}, 0)$.

Notice that the π-vector $(\frac{1}{3}, \frac{1}{3})$ (from (iv) and (v)) is
not extreme. A further reduction is possible in an attempt
to find the extreme covariances. Consider moving a p+1
window <u>along</u> a sequence and evaluating from each position of
the window a contribution to the π-vector. For example for
p = 3 we might have

$$...001[1100]011 ...$$

the contribution from the window [1100] evaluated from the
left hand entry is $\frac{1}{N}(1,0,0)$. However the same contribution
would be obtained from [0011] (reversing the 0's and 1's).
If $\underset{\sim}{x}$ is a (p+1)-tple we call $\bar{x}$ the (p+1)-tple with the 1's
and 0's reversed ($\bar{x}_i \equiv x_i + 1 \bmod 2$). We call $\bar{\underset{\sim}{x}}$ the dual of $\underset{\sim}{x}$.
We say, then, that $(1,x_1,\ldots,x_p)$ and $(0,\bar{x}_1,\ldots,\bar{x}_p)$ are in the
same equivalence class because they both "contribute" to the
same π-vector. That is, given a distribution $\underset{\sim}{P}$ on the
(p+1)-tples the π-vector is

$$
\pi = \sum_{x_1,\ldots,x_p=0,1} (x_1,\ldots,x_p)\{P(1,x_1,\ldots,x_p) + P(0,\bar{x}_1,\ldots,\bar{x}_p)\}
$$

Thus changing all $(0,\bar{x}_1,\ldots,\bar{x}_p)$ in the equations (5) to
$(1,x_1,\ldots,x_p)$ we get a $\frac{1}{2}$ reduction of the original equations.
We may also delete the initial 1 from each (p+1)-tple to get
equations in p-tples. For example the equation (6) becomes

$$
\begin{array}{rcl}
1\ 1\ +\ 0\ 0 & = & 1\ 1\ +\ 1\ 0 \\
1\ 0\ +\ 0\ 1 & = & 0\ 1\ +\ 0\ 0
\end{array}
\qquad (8)
$$

The p-tple $x_1, \ldots, x_p$ now means $P(0, x_1, \ldots, x_p) + P(0, \overline{x}_1, \ldots, \overline{x}_p)$.
We then apply Theorem 1 to the reduced system to identify the
extreme distributions as cycles in the reduced system. For
example the equations (8) have cycles

$$
\begin{aligned}
1\ 1 &\rightarrow (1\ 1\ ) \\
0\ 1 &\rightarrow (0\ 1\ ) \\
0\ 0 &\rightarrow 1\ 0 \rightarrow (0\ 0)
\end{aligned}
\qquad (9)
$$

The reduced system cycles can be traces back to a cycle in the
original system and hence to a periodic sequence. The resulting sequences have the property that they do not (within a
cycle) repeat a p-tple or go through a p-tple and its dual.
The reduced cycles (9) correspond to (i) and (ii) (iii), and
(vi) of equations (7), respectively. The non-extreme
covariance cycles (iv) and (v) have been eliminated by the
reduction; (iv), say, gives the reduced cycle

$$1\ 0 \rightarrow 0\ 1 \rightarrow 0\ 0 \rightarrow \left(1\ 0\right)$$

which is decomposed into

$$1\ 0 \rightarrow 0\ 0 \,\rightarrow (1\ 0) \quad \text{and} \quad 0\ 1 \rightarrow (0\ 1).$$

The original sequence (iv) has 01 and 10 in a cycle.
Theorem 2. If the covariance function $c(1) \ldots c(p)$ (or
$(\pi_1, \ldots, \pi_p)$-vector) of a stationary ergodic binary process is
extreme then it must be obtainable from a strictly periodic
process whose cycle (i) does not repeat a p-tple and (ii) does
not contain a p-tple and its dual.

As a final example in this section consider the case
p = 3. The 16 original equations reduce to

$$
\begin{aligned}
0\ 1\ 1 &+ 1\ 0\ 0 = 0\ 0\ 0 + 0\ 0\ 1 \\
0\ 1\ 0 &+ 1\ 0\ 1 = 0\ 1\ 0 + 0\ 1\ 1 \\
0\ 0\ 1 &+ 1\ 1\ 0 = 1\ 0\ 0 + 1\ 0\ 1 \\
0\ 0\ 0 &+ 1\ 1\ 1 = 1\ 1\ 0 + 1\ 1\ 1
\end{aligned}
$$

The reduced cycles are

$$1 \ 1 \ 1 \to (1 \ 1 \ 1)$$
$$0 \ 1 \ 0 \to (0 \ 1 \ 0)$$
$$0 \ 1 \ 1 \to 0 \ 0 \ 1 \to 1 \ 0 \ 1 \to (0 \ 1 \ 1)$$
$$1 \ 0 \ 0 \to 0 \ 0 \ 1 \to (1 \ 0 \ 0)$$
$$1 \ 0 \ 0 \to 0 \ 0 \ 0 \to 1 \ 1 \ 0 \to (1 \ 0 \ 0)$$
$$0 \ 1 \ 1 \to 0 \ 0 \ 0 \to 1 \ 1 \ 0 \to 1 \ 0 \ 1 \to (0 \ 1 \ 1) \qquad (10)$$

The first five are traced back to the sequences in section 1. However the last gives $\pi = (\frac{1}{2}, \frac{1}{2}, \frac{1}{2})$ and is not extreme. Thus Theorem 2 only gives a necessary condition.

The main technique to find the Π-region is to first use the reduction described here to eliminate as many as possible of the points. The rest are tested by a straight forward method to discover whether a "new" point is in the convex hull of the old points. This is essentially the method adopted by Martins de Carvalho and Clark (1981) but without explicitly describing the reduction of Theorems 1 and 2. They find the extreme covariances up to lag p = 6. (For p = 6 the longest cycle has length 14!)

An important result of Shepp (1967) allows us to explicitly mix together two sequences with π vectors π_1 and π_2 to obtain a new binary sequence with $\pi = (1-\alpha)\pi_1 + \alpha\pi_2$ for any $0 \leq \alpha \leq 1$. This allows us finally to claim that the Π-region is indeed the convex null of the extreme π. The method consists of putting end to end increasingly large segments of either sequence. Note that such a mixed sequence need not be periodic.

An interesting point is the point $\pi = (\frac{1}{k}, \ldots, \frac{1}{k})$ $= (\frac{1}{2}, \frac{1}{2}, \ldots, \frac{1}{2})$ when k = 2. This is clearly obtainable (with probability one) as the realisation of a Bernoulli sequence. However it can also be derived from a full length cyclic error correcting code (also called pseudo-random sequences or de Bruijn sequences). This π-vector is shown in Kiefer and Wynn (1981) to be minimax. That is, it achieves the minimum over the choice of design sequence of

$$\max_{\rho} \ \Phi(c)$$

for all Φ in the class of section 1. It guards against
extreme processes and seems natural when ρ is completely
unknown. We return to pseudo-random schemes in section 4.

4.   HIGHER DIMENSIONS
   The theory of first and second and higher order adjacencies in the last section can be extended to higher dimensions.
We confine ourselves to a straightforward exposition of the
combinatorial theory. The statistical motivation here is less
well developed. One of the problems with the latter is the
general difficulty with defining and elucidating the covariance
structure of auto-regressive processes on multi-dimensional
lattices. We hope to return to a full generalisation of the
statistical theory of Kiefer and Wynn (1981) in a later paper.
Even with the combinatorial theory there seems to be an
explosion of complexity and computer solutions need to be
devised for many of the problems. However the "by hand"
constructions given here point the way and we have been able
to "solve" the first order problem.
   Proceeding as far as possible as in one-dimension we ask
first for a generalisation of the equations (5). Let X_{st} be
a stationary binary process on the infinite 2-dimensional
integer lattice: s,t = 0, ±1,±2,..., X_{st} = 0 or 1. Stationarity requires that for any subsets $\{i_1,...,i_n\}$ and $\{i_1,...,j_n\})$
$(X_{i_r,j_r}, ..., X_{i_n,j_n})$ has the same distribution as
$(X_{i_r+i,j_r+j},...,X_{i_n+i,j_n+j})$ for all integer pairs (i,j).
   Now consider a window of side p+1 containing $(p+1)^2$
entries. Placing this in any position say i,j = 0,...,p we
see that the stationary process induces a distribution $\underset{\sim}{P}$ in
the window. Thus for p = 1 we have a distribution on all 16
2×2 squares:

$$0\ 0 \qquad 1\ 0 \qquad\qquad 1\ 1$$
$$0\ 0\quad ,\quad 0\ 0\quad ,\quad \cdots,\quad 1\ 1 \qquad . \qquad\qquad (11)$$

For example, if the process is completely random:
$\text{Prob}(X_{ij} = 1) = \frac{1}{2}$ independently for all i,j then the
probability of any square (11) is $\frac{1}{16}$.

Proceeding as in the one-dimensional case let a typical $(p+1) \times (p+1)$ square be written X. Partition X into its first column x' and the rest X' thus

$$\underset{\sim}{X} = [x' : X'].$$

Let P(X) be the probability that the induced distribution P assigns to $\underset{\sim}{X}$. Stationarity under horizontal shifts gives us the analogue of equations (5). Namely

(H) $P(\underset{\sim}{X'}) = \underset{x'}{\Sigma} \ P[x' : X'] = \underset{x'}{\Sigma} \ P[X' : x']$

We obtain also under vertical shifts and a horizontal partition

$$\underset{\sim}{X} = \begin{bmatrix} y' \\ \cdots \\ Y' \end{bmatrix} \ ,$$

(V) $P(Y') = \underset{y'}{\Sigma} \ P\begin{bmatrix} y' \\ Y' \end{bmatrix} = \underset{y'}{\Sigma} \ P\begin{bmatrix} Y' \\ y' \end{bmatrix} \ ,$

In (H) (and (V)) (which stand for horizontal and vertical) the summation is over all binary p+1-tples x' (and y').

We can use an analogous notation to that used in (6) etc. dropping the P.   Thus for p = 1 we obtain for (H)

$$\begin{array}{cccc} \frac{00}{00} & + \frac{00}{10} & + \frac{10}{00} & + \frac{10}{10} \end{array} = \begin{array}{cccc} \frac{00}{00} & + \frac{00}{01} & + \frac{01}{00} & + \frac{01}{01} \end{array}$$

$$\begin{array}{cccc} \frac{00}{01} & + \frac{00}{11} & + \frac{10}{01} & + \frac{10}{11} \end{array} = \begin{array}{cccc} \frac{00}{10} & + \frac{00}{11} & + \frac{01}{10} & + \frac{01}{11} \end{array}$$

$$\begin{array}{cccc} \frac{01}{00} & + \frac{01}{10} & + \frac{11}{00} & + \frac{11}{10} \end{array} = \begin{array}{cccc} \frac{10}{00} & + \frac{10}{01} & + \frac{11}{00} & + \frac{11}{01} \end{array}$$

$$\begin{array}{cccc} \frac{01}{01} & + \frac{01}{11} & + \frac{11}{01} & + \frac{11}{11} \end{array} = \begin{array}{cccc} \frac{10}{10} & + \frac{10}{11} & + \frac{11}{10} & + \frac{11}{11} \end{array} \qquad (12)$$

and for (V) the equations with every square replaced by its transpose (we call these (12) (H) and (12) (V)).

Now these two sets of equations (H) and (V) represent necessary conditions for stationarity. We shall investigate distribution $\underset{\sim}{P}$ which satisfy (H) and (V) and are therefore candidates for distributions actually realised under an original process X_{st}. We must emphasize that we have not

proved in general that (H) and (V) are sufficient for a
process on the infinite lattice giving such a $\underset{\sim}{P}$ to exist.
That is (H) and (V) are only <u>local</u> equations. It is quite
conceivable that a particular solution can only be realised
by a distribution on some other 2-dimensional manifold. This
seems to us a profoundly interesting topological problem but
one we shall defer to further research. Fortunately for all
the examples considered here we can <u>exhibit</u> processes on the
planar (or higher dimensional) lattice with the required
solution P. Thus existence will not be a problem.

We shall proceed now largely to examples, and begin with
(12) (H) and (V) for p = 1. The extreme distributions are
found by the direct analogue of Theorem 1 and consist of
(horizontal) cycles which do not visit the LHS of any equation
more than once. That is they must not repeat a vertical
2-tple and are of length ≤ 4.

The one-dimensional theory connecting c(r) with π_r goes
over to the two-dimensional case in the following sense.
Define the values for the four points of the compass:
$\pi = (\pi^N, \pi^E, \pi^{NW}, \pi^{NE})$ then the contribution to π from $P(\underset{\sim}{X})$ where

$$X = \begin{bmatrix} x_{11} & x_{12} \\ x_{21} & x_{22} \end{bmatrix}$$

is

$$P(x)(1+x_{11}+x_{21}, \ 1+x_{11}+x_{12}, \ 1+x_{12}+x_{21}, \ 1+x_{11}+x_{22})$$

where the vector is reduced (mod 2). The contribution of
$\begin{bmatrix} 1 & x_{12} \\ x_{21} & x_{22} \end{bmatrix}$ is the same as that of $\begin{bmatrix} 0 & \bar{x}_{12} \\ \bar{x}_{21} & \bar{x}_{22} \end{bmatrix}$, its dual. Thus
we can form reduced equations by replacing every square in
(12) H which has $x_{11} = 0$ by its dual. We get 4 equations:

$$\begin{matrix} \frac{10}{00} + \frac{10}{10} + \frac{11}{01} + \frac{11}{11} = \frac{10}{10} + \frac{10}{11} + \frac{11}{10} + \frac{11}{11} \end{matrix}$$

(H')

$$\begin{matrix} \frac{11}{00} + \frac{11}{10} + \frac{10}{01} + \frac{10}{11} = \frac{10}{00} + \frac{10}{01} + \frac{11}{00} + \frac{11}{01} \end{matrix}$$

We now need only consider cycles of length 2. They can
easily be listed together with their π-vectors

Cycle π-vector

$\begin{matrix} 11 \\ 11 \end{matrix} \rightarrow \begin{pmatrix} 11 \\ 11 \end{pmatrix}$ $(1, 1, 1, 1)$

$\begin{matrix} 10 \\ 10 \end{matrix} \rightarrow \begin{pmatrix} 10 \\ 10 \end{pmatrix}$ $(1, 0, 0, 0)$

$\begin{matrix} 11 \\ 00 \end{matrix} \rightarrow \begin{pmatrix} 11 \\ 00 \end{pmatrix}$ $(0, 1, 0, 0)$

$\begin{matrix} 10 \\ 01 \end{matrix} \rightarrow \begin{pmatrix} 10 \\ 01 \end{pmatrix}$ $(0, 0, 1, 1)$

$\begin{matrix} 10 \\ 00 \end{matrix} \rightarrow \begin{matrix} 11 \\ 10 \end{matrix} \rightarrow \begin{pmatrix} 10 \\ 00 \end{pmatrix}$ $(\frac{1}{2}, \frac{1}{2}, 0, 1)$

$\begin{matrix} 11 \\ 01 \end{matrix} \rightarrow \begin{matrix} 10 \\ 11 \end{matrix} \rightarrow \begin{pmatrix} 11 \\ 01 \end{pmatrix}$ $(\frac{1}{2}, \frac{1}{2}, 1, 0)$

$\begin{matrix} 10 \\ 00 \end{matrix} \rightarrow \begin{matrix} 10 \\ 11 \end{matrix} \rightarrow \begin{pmatrix} 10 \\ 00 \end{pmatrix}$ $(\frac{1}{2}, \frac{1}{2}, \frac{1}{2}, \frac{1}{2})$

$\begin{matrix} 11 \\ 01 \end{matrix} \rightarrow \begin{matrix} 11 \\ 10 \end{matrix} \rightarrow \begin{pmatrix} 11 \\ 01 \end{pmatrix}$ $(\frac{1}{2}, \frac{1}{2}, \frac{1}{2}, \frac{1}{2})$

Now the cycles for the corresponding vertical reduced equations (V') will give exactly the same set of π-vectors. Thus
the extreme π-vectors for solution (12) (H) and (V) are precisely the first 6 in the above list.

To show existence we need to trace back to a process on
the lattice which gives each π. But we can easily exhibit such
processes they are <u>period patterns</u> in the plane (with a random
position if we want the stationary process). We list below
the standard pattern for each, it being understood that the
patterns are to be repeated indefinitely vertically and hori-
zontally (we exclude dual patterns)

pattern	π-vector
1 1 1 1	$(1,\ 1,\ 1,\ 1)$
1 1 0 0	$(0,\ 1,\ 0,\ 0)$
1 0 1 0	$(1,\ 0,\ 0,\ 0)$
1 0 0 1	$(0,\ 0,\ 1,\ 1)$
1 1 0 0 1 0 0 1 0 0 1 1 0 1 1 0	$(\frac{1}{2},\ \frac{1}{2},\ 0,\ 1)$
1 1 0 0 0 1 1 0 0 0 1 1 1 0 0 1	$(\frac{1}{2},\ \frac{1}{2},\ 1,\ 0)$

A generalisation of the Shepp result on mixing allows us to claim that the full Π-set is the convex hull of the 6 extreme π-vectors. These extreme patterns (processes) do not seem to have been exhibited before.

This whole analysis can be extended to higher dimensions. Consider the case of three dimensions and $p = 1$. Now the method works with little cubes of side 2 and 8 elements $\{x_{ijk}\}$. We will have vertical, horizontal and "sideways" (S) equations obtained by "moving" the cubes along the three axes. Each block of equations will have 16 equations one for each two dimension planar square perpendicular to the axis in question. Even with the "reduction" this still leaves 8 equations with 16 terms on either side. We shall not present the full solution here, rather we shall show how symmetry conditions can be used to simplify the problem; another kind of reduction.

Let us first generalise the "points of the compass"
directions to 3 dimensions. Label a π-coordinate by its
direction measured from (0, 0, 0). That is $\pi(i,j,k)$ is the
coordinate in the direction (0, 0, 0) $\leftrightarrow$ (i,j,k) where
$i,j,k = 0,\pm 1$, $(i,j,k) \neq (0, 0, 0)$, and $\pi(i,j,k) = \pi(-i,-j,-k)$.
Thus there are 13 possible π-coordinates: 3 axial directions,
3 pairs of short (planar) diagonals and four long diagonals.
For m-dimensions there are $\frac{1}{3}(3^m-1)$ coordinates (= 4 when
m = 2). The finite Euclidean symmetry condition is that

$$\pi(i,j,k) = \pi(|i|,|j|,|k|) \tag{13}$$

Another way of considering this is that if i,j,k = 0,1 and
i',j',k' = 0,1 then the π in the direction (i,j,k)
$\leftrightarrow$ (i',j',k') is the same as that (i+r, j+s, k+t)
$\leftrightarrow$ (i'+r, j'+s, k'+t) for any r,s,t = 0,1 where all
coordinates are reduced (mod 2). Thus the π's are invariant
over the finite Euclidean group over GF(2). This reduces the
distinct number of coordinates to 7: 3 axial direction,
3 planar diagonals, and one large diagonal (in general 2^m-1
coordinates).

    This condition leads to a spectacular reduction in the
equations (H)'. We now show that any solution to (H)'
((V') or (S')) with finite Euclidean symmetry <u>can be</u>
<u>decomposed into cycles of maximum length 2</u>. The proof runs
as follows (and applies to any number of dimensions). Consider
a cycle of 2×2×2 cubes in the (H) direction

$$X_1 \to X_2 \to X_3 \ldots \to X_L \to (X_1) \tag{14}$$

Then consider the 2-cycle

$$X_i \to \tilde{X}_i \to (X_i)$$

in the horizontal direction, where $\tilde{X}_i$ is the reflection of X_i
in its leading 2×2 plane. Then we can realise the same
π-vector as (14) by mixing together the 2-cycles with equal
proportions (i = 1,...,L). Furthermore each such 2-cycle
distribution can actually be realised in 3 dimensions by a
pattern which repeats X_i and its reflection $\tilde{X}_i$ alternatively
in every axial directions. Now the contribution of X_i and $\tilde{X}_i$

to a π-vector with finite Euclidean symmetry is the same.
Thus the set of all distributions is just those obtained
from all cubes X_i. The extreme subset is easily found.
Below, then, is listed the extreme π-vectors with Euclidean
symmetry with the corresponding cube (of zeros and ones).
The vertices of the cube are listed in order

(0,0,0), (1,0,0), (0,1,0), (0,0,1), (1,1,0), (1,0,1),
$$(0,1,1), \quad (1,1,1)$$

and the π-vector coordinates also in the corresponding standard order π((0,0,0) → (i,j,k)) in the position of (i,j,k)
in the above list (excluding i = j = k = 0).

Cube	π-vector	
(1,1,1,1,1,1,1,1)	(1,1,1,1,1,1,1)	
(1,1,1,0,1,0,0,0)	(1,1,0,1,0,0,0)	
(1,1,0,1,0,1,0,0)	(1,0,1,0,1,0,0)	
(1,0,1,1,0,0,1,0)	(0,1,1,0,0,1,0)	
(1,1,0,0,0,0,1,1)	(1,0,0,0,0,1,1)	
(1,0,1,0,0,1,0,1)	(0,1,0,0,1,0,1)	
(1,0,0,1,1,0,0,1)	(0,0,1,1,0,0,1)	
(1,0,0,0,1,1,1,0)	(0,0,0,1,1,1,1)	(15)

(Notice that the cubes are the complete 2^3 factorial plus all
$\frac{1}{2}$ orthogonal fractions hinting at a kind of 2^m spatial
factorial theory.)

   We may impose a further symmetry, namely rotation
invariance

$$\pi(i,j,k) = h(i^2 + j^2 + k^2)$$

for some function h. This is called isotropy for spatial
processes. The extreme π under this restriction are quickly
obtained from the Euclidean case. There are just three
distinct coordinates (m in general): one axial, one short-
diagonal and one long-diagonal. We merely average the
coordinates in (15) appropriately. The extreme isotropic
π-vectors are

$$(1, 1, 1), \quad (\tfrac{2}{3}, \tfrac{1}{3}, 0), \quad (\tfrac{1}{3}, \tfrac{1}{3}, 1), \quad (0, 1, 0).$$

Notice that these are <u>not</u> the same set as for p = 3 in one-dimension, $((\frac{1}{2}, 0, \frac{1}{2})$ is missing). These extreme π-vectors (covariance functions) for an isotropic process in 3-dimensions seem to be new. To obtain the actual processes we must mix together the patterns for the Euclidean case (and set $c(r) = 2\pi_r - 1$).

5.   <u>PSEUDO-RANDOM PATTERNS</u>

In Kiefer and Wynn (1981) we mention how full length error correcting codes can be used to generate sequences with $\pi_r = \frac{1}{k}$ (r = 1,...,p) that is the same π-values as would be obtained (with probability one) by an infinite Bernoulli sequence. Briefly we take a primitive polynomial h(x) of degree p+1 over GF (q^m) where q is prime. That is all the elements of the cyclic group of the polynomial field GF (k^p) (where $k = q^m$) can be generated as powers of a solution x (in GF(k^p)) of h(x) = 0. Then take h(x) and use it as a shift register to generate a full length code. Add an additional zero and repeat the code indefinitely.

For example take q = 2, m = 2, k = 4. Label the elements of GF(2^2) : {0, 1, W, W+1}, where we assume that $w^2 + w + 1 = 0$. Then we seek a primitive polynomial with coefficients in GF(2^2) of degree 2 (we take p = 1). We may take $z^2 + wz + w = 0$. The shift register then says $V_t = V_{t-1} + V_{t-2}$ where $V_t = 0,1,W$ or W+1, and arithmetic is over the addition group of GF(2^2). This leads to the code:

0 1 w 1 1 0 w (w+1) w w 0 (w+1)1(w+1)(w+1) . (16)

Add an addition 0 and repeat. We get a first order pseudo-random sequence, every member of GF(2^2) appears next to each other an equal number of times. This can be seen by moving a window of length 2 along the sequence. The process obtained by giving this sequence a random phase shift is locally uncorrelated.

A trivial method of extending the $\pi_r = \frac{1}{k}$ property to higher dimensions is to take the Kronecker sum of the codes with respect to the group GF(q^m) (or any group on the set of q^m elements). That is for two dimensions if the 2-dimension pattern is V_{ij}

$$V_{ij} = V_i + V_j \quad \text{(over GF}(q^m))$$

Using the cyclic group on four elements (16) leads to

```
A A B C B B A C D C C A D B D D
A A B C B B A C D C C A D B D D
B B C D C C B D A D D B A C A A
C C D A D D C A B A A C B D B B
B B C D C C B D A D D B A C A A
B B C D C C B D A D D B A C A A
A A B C B B A C D C C A D B D D
C C D A D D C A B A A C B D B B
D D A B A A D B C B B D C A C C
C C D A D D C A B A A C B D B B
C C D A D D C A B A A C B D B B
A A B C B B A C D C C A D B D D
D D A B A A D B C B B D C A C C
B B C D C C B D A D D B A C A A
D D A B A A D B C B B D C A C C
D D A B A A D B C B B D C A C C
```

This technique does not in general lead to windows which have the 2×2 window property. Not every possible $(2^2)^4$ window "view" appears equally often. The harder problem of extending the window property to higher dimensions has been studied by MacWilliams and Sloane (1976), Van Lint, MacWilliams and Sloane (1979) and Gilbert (1980). Note however that this theory being based on codes usually excludes the zero view (all entries zero) and therefore produces a process with small constant negative autocovariance. The theory does however produce many interesting examples of two-dimensional patterns which simultaneously have the window property for windows of different size, e.g. 4×1, 2×2, 1×4.

Finally we return to the k = 2, p = 1 two-dimensional problem solved in the last section, (13). A two-dimensional pattern with $\pi = (\frac{1}{2}, \frac{1}{2}, \frac{1}{2}, \frac{1}{2})$ is easily seen by taking the Kronecker sum of the code 011(0). Rearranging this gives (repeated indefinitely)

```
1 1 0 0
1 1 0 0
0 0 1 1
0 0 1 1
```

However this does not have the 2×2 window property (there is a predominance of the square $\begin{smallmatrix} 1 & 0 \\ 0 & 1 \end{smallmatrix}$). The following remarkable 4×4 square has the 2×2 window property (including the zero view).

```
1 1 1 0
1 1 0 1
0 1 0 0
1 0 0 0
```

Van Lint et al (1979) report that it is not possible to find a 4×16 array with the 4×1, 2×2 and 1×4 window properties. Note that our square does not have the 2×1 and 1×2 window property.

In Memoriam

The second author would like to express his great sadness at the death of Professor Jack Kiefer. This paper was prepared directly from rough notes written in July 1981.

Acknowledgements

Thanks are due to David Brillinger, Perci Diaconis and Larry Shepp for help with the literature in coding and communication theory. The research was partially supported by National Science Foundation Grant MCS78-25301.

REFERENCES

Gilbert, E.N. (1980). Random colorings of a lattice of squares in the plane. Siam., J., Alg. Disc. Math. 2, 152-159.

Hobby, C. and Ylvisaker, N.D. (1964). Some structure theorems for stationary probability measures on finite state spaces. Ann. Math. Statist., 35, 550-556

Kiefer, J. (1960). Optimum experimental designs, V, with applications to systematic and rotatable designs. Proc. Fourth Berk. Symp., 2, 381-405.

Kiefer, J. and Wynn, H.P. (1980). Optimum balanced block and
   Latin square designs for correlated observations. Ann.
   Statist., 9, 737-757.

Kiefer, J. and Wynn, H.P. (1981). Optimum and minimax exact
   treatment designs for one-dimensional autoregressive
   error processes. Ann. Statist., (submitted).

MacWilliams, F.J. and Sloane, N.J.A. (1976). Pseudo-random
   sequences and arrays. Proceedings I.E.E.E. 64, 1715-1729.

Martin, R.J. (1981). Some aspects of experimental design and
   analysis when errors are correlated. Biometrika,
   (to appear).

Martins de Carvalho, J.L. and Clark, J.M.C. (1981). Charact-
   erising the autocorrelations of binary sequences. Mimeo,
   Imperial College, London.

Masry, E. (1972). On covariance functions of unit processes.
   Siam. J. Appl. Math. 23, 28-33.

Shepp, L.A. (1967). Covariances of unit processes. Proc.
   Working Conference Stochastic Processes, Santa Barbara,
   California.

Williams, R.M. (1952). Experimental designs for serially
   correlated observations. Biometrika, 39, 151-167.

Van Lint, J.H., MacWilliams, F.J. and Sloane, N.J.A. On
   pseudo-random arrays. Siam. J. Appl. Math. 36, 62-72.

Zaman, A. (1981). An approximation theorem for finite
   Markov exchangeability. Technical Report 176,
   Department of Statistics, Stanford University.

Department of Statistics
University of California at Berkeley
Berkeley, CA 94720

Department of Mathematics
Imperial College
South Kensington
London S.W. 7, England

Index